MW01152194

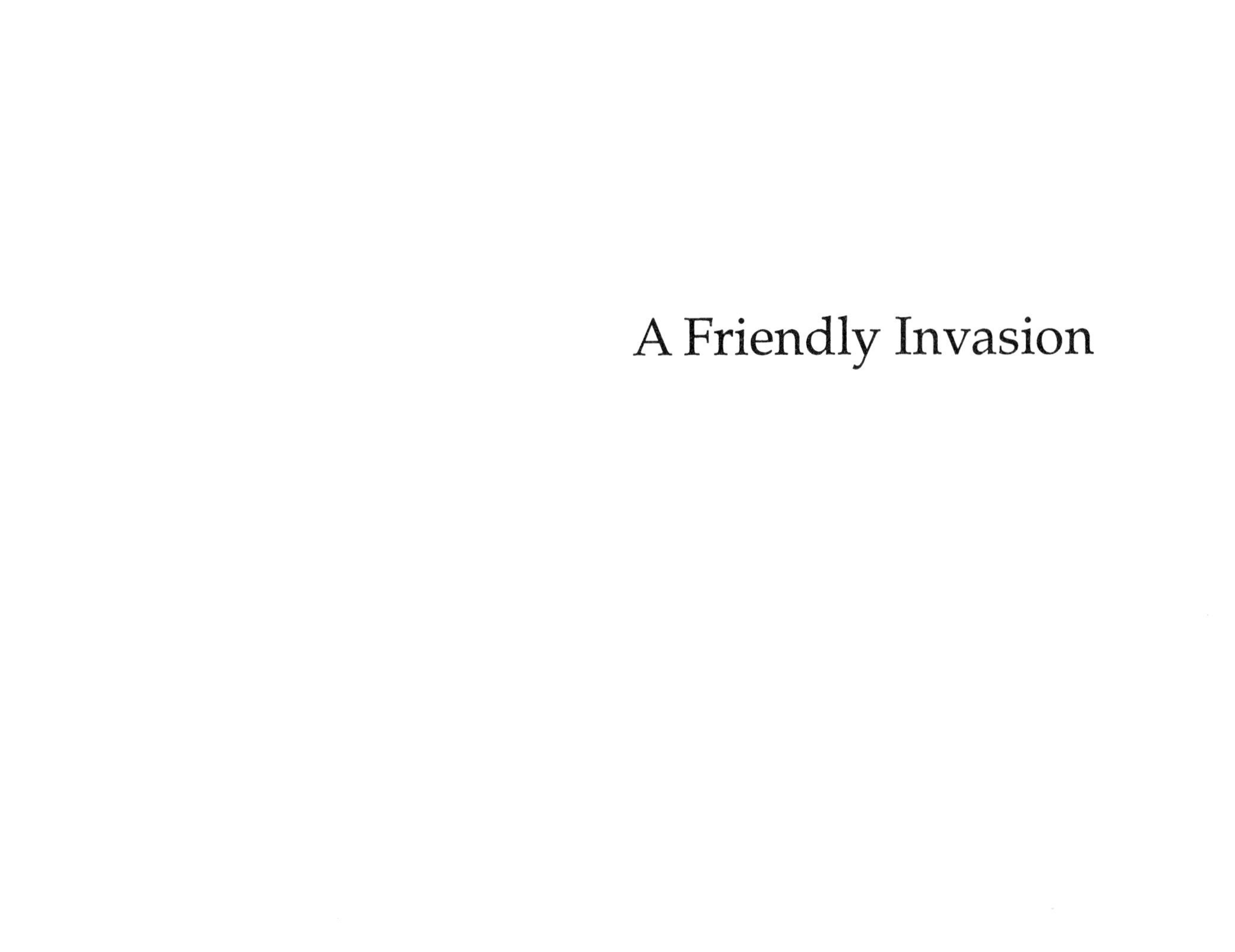

A Friendly Invasion

A Friendly Invasion

The American Military in Newfoundland 1940-1990

John N. Cardoulis

Breakwater
100 Water Street
P.O. Box 2188
St. John's, Newfoundland
A1C 6E6

The Publisher acknowledges the financial support of the Division of Cultural Affairs, Department of Municipal and Provincial Affairs, Government of Newfoundland and Labrador, and of The Canada Council which have helped make this publication possible.

Canadian Cataloguing in Publication Data

Cardoulis, John N.

A friendly invasion

ISBN 0-920911-85-4

1. Military bases, American–Newfoundland–History. 2. United States–Armed Forces–Newfoundland–History. 3. Newfoundland–History, Military. I. Title

UA26.N5C37 1990 355.7'09718 C097542-3

Printed in Canada.
10 9 8 7 6 5 4 3 2 1

To the members of
the United States Army, Navy, Air Force and Marines
and to all the civilian personnel who served and worked on over
sixty United States military installations in Newfoundland and
Labrador, Northern Quebec and Greenland from 1940 to 1990.

Acknowledgements

Many articles, papers and editorials have been written since 1940 on the social and economic impact the American military had on the Newfoundland people. This book may be considered today as a part of Newfoundland history, and viewed by many, in the future, as part of our Newfoundland heritage. No one has ever elaborated on the reasons why the American military came here in 1940, and what they accomplished. The object of this book is not only to tell the story of some of the highlights of the past fifty years, but to substantiate the writings with vintage photographs and documents. The historical events that occurred in Newfoundland and Labrador, which were attributed to the American Forces being here from 1940 onward, can now be placed in their proper perspective and sequence for the benefit of our youth. Over 100,000 American servicemen served here in over sixty locations, and over one-quarter of them married Newfoundland girls. Over 750,000 American servicemen passed through Gander, Stephenville, Argentia, St. John's and Goose Bay during the war years and after. The legacy they left behind them in 1976, when all but the personnel at Argentia Naval Station vacated this province, will never be forgotten; neither will the happy memories of our association with them over the years, and still today. That is why some of their accomplishments and achievements had to be told.

The writing of this story began in 1987, when preparations were first made by the American Legion, Fort Pepperrell Post #9, to participate in the "Great '88 Soiree," and their campaign in 1988 of "Yank! Come Back to Newfoundland." A very special word of thanks is given to the thousands of American and Newfoundland visitors who were interviewed during the "American Military Historical Review in Newfoundland and Labrador," on display at Fort Pepperrell Post #9 for seven months. The inspiration given by the many visitors prompted the writing of this book and the collection of the vintage photographs and documents.

A sincere debt of gratitude is owed to those who contributed photographs, documents, stories of events, research information and incidents of happy and memorable times. A special thanks to over 1,200 American ex-servicemen who served here and wrote letters of encouragement filled with information.

I am most grateful to all those who assisted in my the years of research, particularly the Centre for Newfoundland Studies, Memorial University; Provincial Archives of Newfoundland and Labrador; US Army, US Air Force, US Navy and US Marines; Time/Life Books Inc.; Paul Brown; the Readers Digest Association of Canada; the US Department of State; all members of Fort Pepperrell Post #9; historical records of the American Legion; historical records of US Armed Forces who served here; *Them Days*; Tom Linigar; the Government of Newfoundland and Labrador; Ena Farrell Edwards; John Pappas; Michael Harrington; Norman Dodd; William Slade; Joe Waller; Paul Kavanaugh; the National Archives of Canada; Canadian Red Cross; Lal Parsons; The *Evening Telegram*; Jim Shields; the *Daily News*; the *Western Star*; *Atlantic Advocate*; Walter Davis; Cecil Hutchens; Wallace Furlong; CFS Gander; M.W. Runyan; Herb Wells; William Callahan; Martin Zelenko; Kevin McDonald; Gerry Angel; Jim Andrews; Monder Bray; Frank Joy; Ralph Walker; Joe Murphy; Gary Knight; Bob Jorgensen; Eric Mullett; Towns of Stephenville, Gander, Colinet; Frank Burke; Mrs. Archibald Stacey; Paul O'Neill; Morely Bursey; Stephen Outerbridge; Charlie Warren; Robert Schamper; James Sloan; Mary MacDonald; Tom Godden; Gander Town Library; Arthur Barrett; Leo Kerwin; Mike Renouf; Joe Santomas; William Parrott; David Barron; Kevin Fagan; Robert Docherty; Betty Gibbons; Ray Rieser; Vince Wiltshire; Doug Hiscock; Andy Coady; Bob Nicol; Oscar Boutell; Lt. Com. Tom Rogers; W/O William Rideout; Lt. Kevin Barron; Larry Smith; Al Westland; Peter Steffins; Parks Canada; A.V. Oswald; Marjorie Hendricks; Francis Jalosky; Tom Clancey; Charlie Riddle; Government of Newfoundland and Labrador; Richard Guess; Jerry Pennucci; Russell Englund; Libby McDonald; Phyllis Dunn; Mrs. Francis Yonkin; Department of National Defence; Peter Ledwick; the *Harmoneer*; the *Proppagander*; the *Yank*; Charlie Hoddinott; Len Bunguay; James LaBella; Randy Russell; Carl Lindahl; Frank Butler; C. Falk Photo; Centre for Newfoundland Studies, Memorial University of Newfoundland; the many copies of vintage American newspapers and magazines; and to the numerous other wonderful people who contributed in every way possible. A very special thanks to Mr. Marcel Juteau and Tony Cardoulis for their photography and development of photos and documents; to Quick Print Services Ltd. for assistance in map making and duplication; and to Nora Hughes for her patience and understanding in typing.

Lastly, I must extend my deepest gratitude to my wife and children and friends, who encouraged me to detail every event that could be researched and recorded. They attentively listened to me as I read aloud everything that I wrote and rewrote on each article and photo caption.

John N. Cardoulis
St. John's, Newfoundland
May 1990

Contents

Abbreviations

AACS
Army Air Communication System
AC&W
Aircraft Control and Warning
ADC
Air Defence Command (US)
AFB
Air Force Base
APO
Army Post Office
ASW
Aerial Surveillance and Warning
ATC
Air Transport Command
AWC
Air Warning Company
BW-1
Bluie West 1 (code name)
BW-8
Bluie West 8 (code name)
CAAA
Coast Artillery & Anti-Aircraft
CADAC
Continental Air Defence Command
CFS
Canadian Forces Station
CNT
Canadian National Telecommunications
CONAD
Continental Air Defence (Command)
CTF
Command Task Force
DOT
Department of Transport (Federal)
FBI
Federal Bureau of Investigation
FIS
Fighter Interceptor Squadron
FPO
Fleet Post Office
GCA
Ground Control Approach
LTA
Lighter-Than-Air
M/Sgt.
Master Sargeant
MAC
Military Air Command
MATS
Military Air Transport Service
NASA
National Aeronautical Space Administration
NCO
Non-commissioned officer
NEAC
Northeast Air Command
NORAD
North American Air Defence (Command)
NRCC
National Research Council of Canada
OSI
Office of Special Investigations
PBYS
Amphibious Aircraft Patrol Boat
PFc.
Private First Class
RAF
Royal Air Force (Britain)
RCAF
Royal Canadian Air Force
RCN
Royal Canadian Navy
RCNVR
Royal Canadian Naval Reserve
64th AD
64th Air Division
S/Sgt.
Staff Sargeant
SAC
Strategic Air Command
SCR
Special Control Radar
SONAR
Sound Navigation Ranging
SSO
Special Services Organization
STADAN
Space Tracking & Data Acquisition Network
T-5
Technical Sargeant, 5th Grade
T/Sgt.
Technical Sergeant (USAF)
USAAC
United States Army Air Corps
USAAF
United States Army Air Force
USAF
United States Air Force
USN
United States Navy
USO
United Service Organization
USS
United States Steamship
UST
United States Transport (ship)
V-E
Victory in Europe
V-J
Victory over Japan
VORG
Voice of Radio Gander
VOUG
Voice of United States (Goose Bay)
VOUS
Voice of United States (Pepperrell)
WACS
Women's Army Corps
WO
Warrant Officer
ZI
Zone of Interior

1

Where It All Began

The peace loving nations of the world were shocked when the German army invaded Poland on 1 September 1939. Prior to that both Austria and Czechoslovakia had fallen under political pressure into German hands. World War II had begun. Both the British and French immediately began to prepare for the defence of their countries. British Prime Minister Neville Chamberlain convened with Parliament, and on Sunday, 3 September 1939, Britain and France declared war on Germany. Conscription was the order of the day in both France and Britain. Other smaller countries in Europe, who feared a German invasion, declared their aggression against Germany and, together with the English and French, began to build up an allied fighting force. England formed the British Expeditionary Force, and four divisions were sent to France to assist the Allied Forces in their efforts to secure the French-German border. Britain also placed a call to all colonies in the British Empire to assist where they could, in the impending and inevitable war with Germany. Canada declared war on Germany on 10 September 1939.

In England, the Lord of the Admiralty, Winston Churchill, had been called to power on 10 May 1940 as the new Prime Minister of England. The same day the Germans invaded Holland, Belgium and Luxembourg. The war had finally come to Western Europe. Part of the Allied Forces in Northern France began an immediate advancement toward Belgium to engage the German army. They were, however, unable to cope with the continuous attacks by the German *Luftwaffe* and *Panzer* tank divisions, and the rapidly advancing German troops to the north and south. They gradually pushed the Allies into a pocket at Dunkirk in Northern France. Holland fell to the Germans on 15 May 1940 and Belgium on 28 May. At Dunkirk, one of the most daring and historic rescues was undertaken by the British to get the trapped Allied soldiers off the beaches. Thousands of ships and boats of every description from small motor boats, pleasure craft, fishing vessels, cargo boats, barges to Royal Navy destroyers, sailed from the English coast to Dunkirk to participate in the rescue. The operations began on 26 May, and by 6 June over 330,000 soldiers of the British Expeditionary Force, as well as French and Belgian troops, were safely transported to England.

Prior to 1940, the American people were very reluctant to involve themselves in the European war. The United States Neutrality Act was amended in 1939, after the German invasion of Poland, to allow the shipment of arms to combatants. Germany was excluded from the amendment. As the war escalated in 1940, and particularly after the initial bombing of London, the American attitude changed. Some US-made Lockheed bombers were shipped to England by sea. However, because of the German U-boat manoeuvres, many did not arrive. After Prime Minister Churchill took office in May 1940, he contacted his old friend Franklin D. Roosevelt, President of the United States, through the Prime Minister of Canada, William Lyon MacKenzie King, and requested assistance in Britain's defence against the German forces. The assistance was to be mainly in the form of military ships and aircraft. Because of the American Neutrality Act, the President could not honour his request, for fear America would automatically involve itself in the war. Negotiations were started immediately between the USA and Canada on behalf of Britain, and a way was found to assist. This was in the form of a Lend Lease of war materials. On 22 August 1940 the United States agreed to supply fifty US Naval destroyers to Britain in exchange for ninety-nine year leases on British territories in Newfoundland, Bermuda and the West Indies. These territories would be required to establish United States military bases. Britain agreed, however the leased bases in Newfoundland and Bermuda would be free. Only the leased land in the West Indies would be subject to the exchange for fifty US Naval destroyers. The agreement was not signed until 27 March 1941 in London. The President of the United States set up a Canada-United States Permanent Joint Board on Defence. The first meeting was held in Ottawa on 26 August 1940. President Roosevelt further agreed to send arms and other military supplies, including aircraft, to Britain. Because the US Neutrality Act prohibited war planes from being flown out of the United States, these bombers were delivered to the US-Canadian border and towed across into Canada. This procedure, of course, changed after 14 December 1941. After the fall of France and the bombing and impending German invasion of England, the American people were then cognizant of the fact that, if the British Isles were conquered, nothing would stop Hitler from extending the war to the Western Hemisphere. The acquisition of land for US military bases, especially in Newfoundland, would thus strengthen the overall defence of

North America. Land for bases was also acquired by the United States in Greenland, Iceland and Northern Canada.

Newfoundland, then a quiet, peaceful British colony, was considered as the most strategic location; a stepping stone between North America and Britain in the defence of the Western Hemisphere. It was the only territory in North America closest to England. Newfoundland was the first to experience the effects of war in North America when, on 5 September 1939, a German merchant ship, the *Cristoph Doornum*, was seized in Botwood and its crew were made prisoners of war. They were transported to St. John's by train on 6 September and a temporary confinement was set up in the YMCA building in St. John's. The prisoners of war were later moved to a German Internment Camp at Pleasantville. Newfoundland was also the first British colony to declare war on Germany, institute rationing, blackouts, and restrictions on wireless telephone and radio stations. All outgoing and foreign mail and all telegrams were censored, and all alien personnel registered and placed under surveillance. Newfoundland was also the first British colony to be attacked by a U-boat. Part of the role played by Newfoundland in the Second World War in the defence of North America is contained in the following chapters of this book. The Royal Newfoundland Regiment was called to service in 1939. In World War II, Newfoundlanders made up two Artillery Regiments of the British Army: the 166th Newfoundland Field Regiment, Royal Artillery (formerly the 57th Heavy Newfoundland Regiment) which served in the United Kingdom, North African and Italian Theatres; and the 59th Newfoundland Heavy Regiment, which fought in the North-West European campaign. The 125th (Newfoundland) Squadron of the Royal Air Force was a night fighter squadron which engaged in helping to defend Britain's coast. Over 3,500 Newfoundlanders served in the Royal Navy and over 700 in the Canadian Navy. By 4 December 1940 over 2,200 recruits for the Royal Navy sailed overseas from Maritime ports and from St. John's. More than 3,500 Newfoundlanders went overseas to serve in the Newfoundland Overseas Forestry Unit and nine hundred Newfoundlanders were killed in action during the Second World War.

While the Royal Air Force was engaging itself in the Battle of Britain, the Royal Navy was involved in many skirmishes in the North Sea and in the North Atlantic. The German *Luftwaffe* and German U-boats preyed upon all shipping to the British Isles. The Royal Navy fought desperately to maintain its supremacy to rule on the sea. The shortage of destroyers and other vessels in early 1940, compounded by the U-boat manoeuvres, caused great concern throughout the British Navy. Toward the later part of 1940, new fighting ships were being launched and more ships from Canada and the United States began to strengthen the Royal Navy. The German Navy concentrated its efforts on destroying all the merchant shipping it could and thus prevent England from receiving much needed war materials and supplies. Furthermore, they were after troop carriers; like snakes in the dark, ready to strike at any time.

The sinking of Allied vessels in the later part of 1940 was considered a great loss, and the conditions escalated as the German U-boat menace increased. Over 150,000 tons of Allied merchant shipping was sunk in February 1941, and over 240,000 tons by 15 March. There were also high loss rates in April and May, which finally mounted to three or four ships sunk each day. Because of bad weather during the first three months of 1941, the German *Luftwaffe* was grounded most of the time, thus allowing the Royal Navy to concentrate more readily on the German submarines. For every U-boat sunk, Germany would replace it with eight more.

On 22 June 1941 Hitler surprised everyone by attacking Russia. This action took the pressure off Britain, as many of the German divisions that had assembled in France for "Operation Sea Lion" were now being transferred to the Russian front. During the same month, the Newfoundland Escort Naval Force was established under the command of Canadian Commodore Murray. The Newfoundland force was made up of many Canadian and Royal Navy vessels as well as others from Allied nations. Commodore Murray's jurisdiction extended to all shipping in and out of St. John's and in Newfoundland coastal waters. While this was happening, over twenty-five German U-boats were on patrol off the south coast of Greenland and east of Newfoundland. The Battle of the Atlantic was well under way by June 1941. The huge US Naval base at Argentia played a major role in protecting convoy ships to Britain, and in tracking down German U-boats. All types of Allied naval fighting ships operated out of Argentia. By the latter part of 1943, the attacks by German submarines were starting to ease off, due to constant patrols by air and sea. The Canadian Navy escorted over 25,000 merchant ships from North America to Britain and back during the war years. By 1942 the German Navy had placed its full concentration of its vessel power in submarines; it would rendezvous in packs in the Atlantic, most times out of range of aircraft from Newfoundland and other Allied bases in the USA, Canada, Greenland, Iceland, Bermuda and Britain. During the winter of 1942-43 there were over 100 U-boats in the Atlantic. Many were patrolling the eastern seaboard of the United States. Allied losses were tremendous; in February 1943 over twenty ships, totalling over 180,000 tons, were sunk. Over eighty ships were sunk in March, and only a few German U-boats were destroyed. These huge losses continued until the middle of 1943 and then tapered off. The use of radar by the Allies, bigger and more modern ships and more long-range aircraft patrols caused the Battle of the Atlantic to turn in favour of the Allies by the beginning of 1944.

Honolulu Star-Bulletin 1st EXTRA

HONOLULU, TERRITORY OF HAWAII, U. S. A., SUNDAY, DECEMBER 7, 1941 ★ PRICE FIVE CENTS

WAR!

(Associated Press by Transpacific Telephone)

SAN FRANCISCO, Dec. 7.—President Roosevelt announced this morning that Japanese planes had attacked Manila and Pearl Harbor.

OAHU BOMBED BY JAPANESE PLANES

SIX KNOWN DEAD, 21 INJURED, AT EMERGENCY HOSPITAL

Attack Made On Island's Defense Areas

By UNITED PRESS

WASHINGTON, Dec. 7. —Text of a White House announcement detailing the attack on the Hawaiian islands is:

"The Japanese attacked Pearl Harbor from the air and all naval and military activities on the island of Oahu, principal American base in the Hawaiian islands."

Oahu was attacked at 7:55 this morning by Japanese planes.

The Rising Sun, emblem of Japan, was seen on plane wing tips.

Wave after wave of bombers streamed through the clouded morning sky from the southwest and flung their missiles on a city resting in peaceful Sabbath calm.

According to an unconfirmed report received at the governor's office, the Japanese force that attacked Oahu reached island waters aboard two small airplane carriers.

It was also reported that at the governor's office either an attempt had been made to bomb the USS Lexington, or that it had been bombed.

CITY IN UPROAR

Within 10 minutes the city was in an uproar. As bombs fell in many parts of the city, and in defense areas the defenders of the islands went into quick action.

Army intelligence officers at Ft. Shafter announced officially shortly after 9 a. m. the fact of the bombardment by an enemy but long previous army and navy had taken immediate measures in defense.

"Oahu is under a sporadic air raid," the announcement said.

"Civilians are ordered to stay off the streets until further notice."

CIVILIANS ORDERED OFF STREETS

The army has ordered that all civilians stay off the streets and highways and not use telephones.

Evidence that the Japanese attack has registered some hits was shown by three billowing pillars of smoke in the Pearl Harbor and Hickam field area.

All navy personnel and civilian defense workers, with the exception of women, have been ordered to duty at Pearl Harbor.

The Pearl Harbor highway was immediately a mass of racing cars.

A trickling stream of injured people began pouring into the city emergency hospital a few minutes after the bombardment started.

Thousands of telephone calls almost swamped the Mutual Telephone Co., which put extra operators on duty.

At The Star-Bulletin office the phone calls deluged the single operator and it was impossible for this newspaper, for sometime, to handle the flood of calls. Here also an emergency operator was called.

HOUR OF ATTACK—7:55 A. M.

An official army report from department headquarters, made public shortly before 11, is that the first attack was at 7:55 a. m.

Witnesses said they saw at least 50 airplanes over Pearl Harbor.

The attack centered in the Pearl Harbor, Army authorities said:

"The rising sun was seen on the wing tips of the airplanes."

Although martial law has not been declared officially, the city of Honolulu was operating under M-Day conditions.

It is reliably reported that enemy objectives under attack were Wheeler field, Hickam field, Kaneohe bay and naval air station and Pearl Harbor.

Some enemy planes were reported shot down.

The body of the pilot was seen in a plane burning at Wahiawa.

Oahu appeared to be taking calmly after the first uproar of queries.

ANTIAIRCRAFT GUNS IN ACTION

First indication of the raid came shortly before 8 this morning when antiaircraft guns around Pearl Harbor began sending up a thunderous barrage.

At the same time a vast cloud of black smoke arose from the naval base and also from Hickam field where flames could be seen.

BOMB NEAR GOVERNOR'S MANSION

Shortly before 9:30 a bomb fell near Washington Place, the residence of the governor. Governor Poindexter and Secretary Charles M. Hite were there.

It was reported that the bomb killed an unidentified Chinese man across the street in front of the Schuman Carriage Co. where windows were broken.

C. E. Daniels, a welder, found a fragment of shell or bomb at South and Queen Sts. which he brought into the City Hall. This fragment weighed about a pound.

At 10:05 a. m. today Governor Poindexter telephoned to The Star-Bulletin announcing he has declared a state of emergency for the entire territory.

He announced that Edouard L. Doty, executive secretary of the major disaster council, has been appointed director under the M-Day law's provisions.

Governor Poindexter urged all residents of Honolulu to remain off the street, and the people of the territory to remain calm.

Mr. Doty reported that all major disaster council wardens and medical units were on duty within a half hour of the time the alarm was given.

Workers employed at Pearl Harbor were ordered at 10:10 a. m. not to report at Pearl Harbor.

The mayor's major disaster council was to meet at the city hall at about 10:30 this morning.

At least two Japanese planes were reported at Hawaiian department headquarters to have been shot down.

One of the planes was shot down at Ft. Kamehameha and the other back of the Wa-

Turn to Page 2, Column 1

Hundreds See City Bombed

Hundreds of Honolulans who hurried to the top of Punchbowl soon after bombs began to fall, saw spread out before them the whole panorama of surprise attack and defense.

Far off over Pearl Harbor the white sky was polka-dotted with anti-aircraft smoke.

Rolling away from the navy base were billowing clouds of ugly black smoke. Sometimes a burst of flame reddened the black sources of the smoke.

Out from the silver surfaced mouth of the harbor a flotilla of destroyers streamed to battle smoke pouring from their stacks.

Turn to Page 2, Column 3

Names of Dead and Injured

The city emergency hospital reported at 10:30 a list of 6 killed and 21 injured.

The complete list will be carried later. Here is a partial list:

Peter Lopez, 34, of [illegible] St., was reported at 9:30 a. m. to be in serious condition from wounds in the upper abdomen.

Bernice Gouveia, 11, 1708 Kalihi St., is suffering from a mangled thigh, lacerations on the right leg and left arm.

A Portuguese girl, unidentified, 16 years old, died on arrival from puncture wounds.

Another victim who died on arrival was Frank Ohashi, 20, 1708 [illegible] St., from puncture wounds in the chest.

Cecelia Broadly, 30, [illegible] gardens, was released from the hospital after treatment for lacerations.

Three were reported injured and one reported killed from the bomb that fell at Fort and School Sts.

Schools Closed

All schools on Oahu, both public and private, will remain closed until further notice. Edouard L. Doty, territorial director of civilian defense, announced at 11 a. m. today. This does not apply elsewhere in the territory.

Editorial

HAWAII MEETS THE CRISIS

Honolulu and Hawaii will meet the emergency of war today as Honolulu and Hawaii have met emergencies in the past—coolly, calmly and with immediate and complete support of the officials, officers and troops who are in charge.

Governor Poindexter and the army and navy leaders have called upon the public to remain calm; for civilians who have no essential business on the streets to stay off; and for every man and woman to do his duty.

That request, coupled with the measures promptly taken to meet the situation that has suddenly and terribly developed, will be needed.

Hawaii will do its part—as a loyal American territory.

In this crisis, every difference of race, creed and color will be submerged in the one desire and determination to play the part that Americans always play in crisis.

BULLETIN

Additional Star-Bulletin extras today will cover the latest developments in this war move.

"The Day of Infamy"

The Japanese attacked Pearl Harbor on 7 December 1941, "The Day of Infamy." The United States of America then declared war on Japan. Germany and Italy declared war on the United States on 14 December 1941. All the restrictions of the US Neutrality Act were immediately lifted and America started to ship war materials, airplanes and ships to Britain and the Far East. The American Forces started an offensive in two strategic areas as of 7 December. To crush Japan, the US War Department knew it first had to help England to defeat Hitler's Germany. The United States Eighth Air Force was established in England and grew into a mighty fighting command. It was engaged in the bombing of Germany and German occupied countries. The US military forces in the Far East were heavily engaged in their fight against the Japanese. The Allied forces invaded North Africa in 1942, Sicily in 1943, and Normandy in 1944. The Russians had held off the German invasion of their country and were beginning to drive the German army back. The Allies liberated German occupied countries and were soon converging upon Germany itself. The end of the European War seemed soon to becoming a reality.

While the United States military bases in Newfoundland had a specific mission of defence prior to 7 December 1941, the overall mission of each facility was now directed toward support in fighting the war, as well as the continued defence of the North Atlantic. Newfoundland, and its territory, Labrador, was once again designated as a vital link in the success of winning the war. The Atlantic Ferry Organization, flying thousands of planes from the United States and Canada to Britain, operated out of Gander and Goose Bay. The huge air movement of troops and war supplies to England operated through Ernest Harmon Air Base in Stephenville. The Newfoundland Escort Service operated out of St. John's and Argentia. The US and Canadian defence of Newfoundland and Labrador, in their establishment of Coast Artillery and Anti-Aircraft Batteries throughout Newfoundland, was accelerated. The US Infantry and Canadian Army policed and secured vital installations.

The Allies continued to liberate country after country in Europe, until Germany was defeated on 8 May 1945. The attention of the Allies was then directed to the Far East, to assist the Americans in the fight with the Japanese. The Japanese losses in 1944-45 and the first atomic bomb drops on Hiroshima on 6 August, and Nagasaki on 9 August, caused the defeat of the Japanese and their surrender on 14 August 1945. The Second World War was over.

The close association between the American military and the Newfoundland people cannot be overemphasized. When the young US soldiers and sailors arrived here in 1940-41, and particularly throughout the war years up to 1945, they were lonesome and bewildered. They left their families and loved ones behind in the

United States. Many never knew that the Island of Newfoundland existed. Those who arrived on the UST *Edmund B. Alexander* were not told where they were going when they left the Port Authority in New York on 15 January 1941. It wasn't until they were off the Newfoundland coast, waiting for the weather to clear before sailing into St. John's, that they were advised they were to go ashore on the Island of Newfoundland and establish bases there and help to defend the Island in case of enemy attack. One can imagine their desperation while aboard the *Alexander* for three days, in a rough sea and snowstorm, catching a glimpse every so often of the rugged Newfoundland coastline. America was not at war at that time, but there was an air of uncertainty in the atmosphere. They knew nothing of the Newfoundland people, or the country.

On the other hand, the Newfoundland people, already shadowed by the Second World War, were hesitant to accept foreigners on their soil. In the first instance, many were bitter over the fact that the Anglo-American Agreement on ninety-nine year leases for Americans was initially authorized by England without the knowledge and consent of the Government of Newfoundland or its people. As the *Alexander* sailed near the coast and while it was offshore waiting to be brought into St. John's Harbour, the hospitality of the Newfoundlanders overcame some of the resentment or fears they had. The reception given the Americans as they sailed through the Narrows was memorable. From that day on, the American military men were accepted as our own; the Americans accepted the extended welcome with much gratitude.

The United States military forces in Newfoundland and Labrador jumped from 2,000 prior to 7 December 1941 to over 20,000 by the end of 1942. By the war's end in 1945, over 100,000 US military troops were stationed in Newfoundland and Labrador. During the period of 1942 to 1946, over 45,000 aircraft, over 10,000 ships and more than 750,000 US military crews and passengers passed through Harmon Field, Gander, Goose Bay, Argentia and Torbay. Nowhere in North America was there such a diversity of military activity from 1940 to 1946. The US and Canadian uniformed soldiers and sailors were as common on the streets of cities and towns in Newfoundland and Labrador as the populace. Britain's oldest colony played a most important role in the Second World War.

On 2 July 1945, almost two months after V-E Day, the Newfoundland Base Commander, Brigadier General S.M. Connell, invited the general public of St. John's to an open house at Fort Pepperrell, prior to the first departure of US Army troops back to the United States. The Commissioner for Defence, the Honourable H.A. Winter, was the guest of honour and, following a review and march past, gave a short address, which was followed by remarks by Brigadier-General Connell. Both emphasized the good will and

No. 1.

The Marquess of Lothian to Mr. Cordell Hull

WASHINGTON, *September 2, 1940.*

SIR,

I HAVE the honour, under instructions from His Majesty's Principal Secretary of State for Foreign Affairs, to inform you that in view of the friendly and sympathetic interest of His Majesty's Government in the United Kingdom in the national security of the United States and their desire to strengthen the ability of the United States to co-operate effectively with the other nations of the Americas in the defence of the Western Hemisphere, His Majesty's Government will secure the grant to the Government of the United States, freely and without consideration, of the lease for immediate establishment and use of Naval and Air bases and facilities for entrance thereto and the operation and protection thereof, on the Avalon Peninsula and on the Southern coast of Newfoundland, and on the East coast and on the Great Bay of Bermuda.

Furthermore, in view of the above and in view of the desire of the United States to acquire additional Air and Naval bases in the Caribbean and in British Guiana, and without endeavouring to place a monetary or commercial value upon the many tangible and intangible rights and properties involved, His Majesty's Government will make available to the United States for immediate establishment and use Naval and Air bases and facilities for entrance thereto and the operation and protection thereof, on the Eastern side of the Bahamas, the Southern coast of Jamaica, the Western coast of St. Lucia, the West coast of Trinidad in the Gulf of Paria, in the Island of Antigua, and in British Guiana within fifty miles of Georgetown, in exchange for Naval and Military equipment and material which the United States Government will transfer to His Majesty's Government.

All of the bases and facilities referred to in the preceding paragraphs will be leased to the United States for a period of ninety-nine years free from all rent and charges other than such compensation to be mutually agreed on to be paid by the United States in order to compensate the owners of private property for loss by expropriation or damage arising out of the establishment of the bases and facilities in question.

His Majesty's Government in the leases to be agreed upon will grant to the United States for the period of the leases all the rights, power and authority within the bases leased, and within the limits of the territorial waters and air spaces adjacent to or in the vicinity of such bases, necessary to provide access to and defence of such bases and appropriate provisions for their control.

Without prejudice to the above-mentioned rights of the United States authorities and their jurisdiction within the leased areas, the adjustment and reconciliation between the jurisdiction of the authorities of the United States within these areas and the jurisdiction of the authorities of the territories in which these areas are situated shall be determined by common agreement.

The exact location and bounds of the aforesaid bases, the necessary seaward, coast and anti-aircraft defences, the location of sufficient military garrisons, stores and other necessary auxiliary facilities shall be determined by common agreement.

His Majesty's Government are prepared to designate immediately experts to meet with experts of the United States for these purposes. Should these experts be unable to agree in any particular situation except in the case of Newfoundland and Bermuda, the matter shall be settled by the Secretary of State of the United States and His Majesty's Secretary of State for Foreign Affairs.

I have, &c.

LOTHIAN.

No. 2.

Mr. Cordell Hull to the Marquess of Lothian.

WASHINGTON, *September 2, 1940.*

EXCELLENCY,

I HAVE received your note of 2nd September, 1940, of which the text is as follows:—

[As in No. 1.][1]

I am directed by the President to reply to your note as follows:—

"The Government of the United States appreciates the declarations and the generous action of His Majesty's Government, as contained in your communications, which are destined to enhance the national security of the United States and greatly to strengthen its ability to co-operate effectively with the other nations of the Americas in the defence of the Western Hemisphere. It therefore gladly accepts the proposals.

[1] These brackets and the brackets on the following pages appear on the certified copy of the agreement.—EDITOR.

"The Government of the United States will immediately designate experts to meet with experts designated by His Majesty's Government to determine upon the exact location of the Naval and Air bases mentioned in your communication under acknowledgment.

"In consideration of the declarations above quoted, the Government of the United States will immediately transfer to His Majesty's Government fifty United States Navy Destroyers generally referred to as the twelve-hundred ton type."

Accept, &c.

CORDELL HULL.

Exchange of Notes in reference to the Anglo-American Agreement on Lend-Lease of Bases in Newfoundland, 2 September 1940. From The American Bases Act.

feelings between the people of Newfoundland and the American Forces at Fort Pepperrell. In a further address to the American soldiers, Lt. Col. Timothy J. Ryan summarized the US mission here in Newfoundland, and the reasons for granting bases to America. He said, "Your job has been a vital one to the war effort, and although you have not heard shell bursts nor seen your buddies die beside you, it is true that you have contributed greatly. When this base was constructed it looked as though the enemy would overrun Britain. They already had weather stations in Iceland and Greenland. Their submarines were running practically unchecked in the North Atlantic. The actual and immediate danger to Canada and the United States was very real indeed. A base from which we could protect our eastern cities and factories was number one priority in our defence plans. The answer to that was Newfoundland. This has become a clenched fist challenging the Germans to come any further. The Germans were driven from Iceland and their stations destroyed on Greenland after the Newfoundland base was secure. While you were here and seemed to be inactive, your very presence was helping to keep the European war confined to Europe."

Everyone looked toward peace and the reduction of each country's military might after V-J Day in August 1945. However, then came the Berlin Airlift, the Korean War, the Vietnam War and the Cold War with Russia. Newfoundland's military position changed considerably from 1950 to 1960, primarily in the defence of North America from Russian missiles and aircraft. Thus, the strategic position of Newfoundland and Labrador remained important, and still does, after fifty years. All of the US military bases and installations have closed except Argentia Naval Station. Two major airports, at Gander and Goose Bay, still maintain a small contingent of US military personnel. What is left of the US bases and sites, besides structures and facilities, are great memories of which the middle-aged and senior citizens of Newfoundland and Labrador will always be proud. Many of the younger generation have yet to learn the importance of their province during the war and afterwards.

September 1990 is the fiftieth anniversary of the American military in Newfoundland. It was in September that the first American forces arrived in St. John's on the USS *St. Lewis*. The first US flag was raised at Argentia on 13 February 1941. At Camp Alexander the first flag was raised on 20 May 1941. The first flag ceremony at Fort Pepperrell was on 1 December 1941, and the last on 11 August 1961. The respect paid to the US flag over the years by Newfoundlanders was clearly demonstrated each day at 8:00 a.m. when the flag was raised, and at 5:00 p.m. when it was taken down. Newfoundland civilians, along with their American friends on each US installation, stood at attention for the three-minute ceremony, twice each day.

The history compiled in the following chapters of this book will hopefully bring back a lot of wonderful memories to many, and cause others to pause in astonishment at the US military activity in Newfoundland for half a century.

This fifty-year-old story, as part of our Newfoundland heritage, had to be told.

These two photos depict familiar sights to the American soldiers and sailors in 1943 and 1944. This building, the old and majestic Newfoundland Hotel, was replaced with a new and more modern hotel in 1985 and renamed Hotel Newfoundland.

Water Street, St. John's 1944, showing Dicks & Co., Bowring Brothers and Ayre & Sons.

2

Argentia Naval Air Station

Argentia, located in Placentia Bay, was a small fishing village in 1940, with 500 people living in 114 buildings. Up to that time, the effects of the war were the furthest thing from the minds of these quiet, happy people. Family names, to list a few, were the McGraths, O'Learys, Roches and the Hunts. The sea provided them their living. Their connections with the rest of Newfoundland were their fishing schooners and the Newfoundland Railway, which operated a Placentia Branch train from St. John's. The Newfoundland Railway also operated a coastal steamer that regularly entered Argentia to load and unload passengers and freight at the Newfoundland Railway wharf.

The following is an excerpt from a Roman Catholic church history record of Argentia, compiled by a US Naval historian.

> Father Adrian John Dee became parish priest [of Argentia] in 1922, and it fell to his lot to see his parish uprooted by the construction of the United States Naval Base.
>
> When England got its fifty over-age destroyers from the United States in 1940, activity started immediately in Argentia. After the rumours became facts, it dawned on the people that they would have to be uprooted, their homes torn down and they would have to settle elsewhere. This movement would include the church, priest's house, school and hall, and three cemeteries.
>
> The first lease for the land was not signed until June 14, 1941. A year later on the same date, a supplementary lease was signed. This meant that the United States controlled a large amount of land in Newfoundland until June 14, 2041.
>
> On March 31, 1941, Father Dee was informed by the Commissioner for Public Works, Sir Wilfred Woods, that the Navy would require the area occupied by the church buildings by May 1, 1941. Father Dee had planned a temporary move to Marquise. The school at Marquise was to be rearranged as a

> residence; a temporary church was to be built, and storage space provided. Father Dee expressed the wish to pay for the church. On April 15, Father Dee was informed that the Marquise area could not be used because the US Army was moving in there (Fort McAndrew).
>
> So then began the trauma, the sorrow and the frustrations of moving from their homes to where they did not know. Father Dee had set himself the goal of keeping his parish intact. With this object in mind, he worked untiringly with his people and with the Commission of Government.

Subsequent to the approval of the Anglo-American Lend-Lease Agreement on 2 September 1940, the American military determined where their facilities would be built. Unnoticed by many, some ground survey work was started early in September 1940 at all locations, plus Signal Hill and other areas where US Coast Artillery was to be stationed and US Infantry look-out towers were to be built. Agreements between the Canadian and US military were also finalized, so there would be a clear-cut understanding of the areas of jurisdiction and responsibility of all.

Taken around 1941, this photo shows some of the dwellings at Argentia. Part of the Newfoundland Railway wharf can be seen, centre left. The large wooden structures to the right are temporary barrack-type buildings for civilian and military personnel.

The first official US military and US civilian engineering personnel to arrive in Newfoundland were on the USS *Bowditch*, which arrived at Argentia on 13 October 1940. These personnel did extensive hydrographic and land surveys at Argentia and at several other proposed military locations. The second US ship to arrive at Argentia was the USS *Richard Peck* on 18 January 1941, carrying 1,500 US civilian construction and engineering personnel. This ship remained tied up at Argentia for two years, and served as the office and living quarters for the US civilians until February 1943. There were also as many as 15,000 local Newfoundland workers employed at that time for construction of the Naval base. Those who did not have their own homes nearby, or who could not stay with relations or friends, lived on schooners anchored in the bay. All food, construction materials and construction equipment were brought in by ship from the United States and unloaded at the nearby Newfoundland Railway wharf. The priority construction projects at the time were to dredge the harbour, to divert a water source to the base and to construct runways.

The first all-US military troops arrived in Newfoundland on 25 January 1941, and consisted of 120 US Marines. They were assigned security and police duties on the construction sites and they were quartered on the *Peck*. Many of the existing buildings in the village were used for offices and officers' quarters. The old school served as an office for the Commandant, aerological unit, harbour entrance contact and government post office. The residents of the village, as well as three cemeteries, were relocated to surrounding areas.

Little Placentia (Argentia), 1940-41. The large buildings in the background are part of the temporary construction .

The spot just ahead of the USS Richard Peck *is the closest point to where the Marines came ashore in January 1941 and raised the American flag. A monument to commemorate the landing was erected here. This photo was taken on 6 January 1941. Note the progress of contruction up to this point in time.*

The Anglo-American agreement for the ninety-nine year lease was implemented by enactment on 11 June 1941 by the Newfoundland Government's statute, known as "The American Bases Act." The first specific reference to Argentia is made in sub-schedule "B" of the Act, which describes a lot of approximately 3,392 acres of land near Placentia. This area was later to be occupied by the US Navy and the US Army at Fort McAndrew. The first construction contract for the Argentia base was awarded to Fuller Construction Company and was signed on 28 January 1941. It provided for a Naval Air Station to be erected at an estimated cost of $9,424,000. Additional contracts were awarded during 1941 to 1944. In October 1944 the overall value of the US Naval base, as assessed by Naval Command in the USA, was $53,200,000. The US Congress recognized Argentia as the most expensive military installation anywhere outside the continental USA during World War Two.

The USS *Abemarles*, a sea plane tender, arrived in Argentia 15 May 1941 to establish a base for patrol and support operations. Preparations were made for the arrival of Patrol Squadron 52, the first US squadron to fly along the Atlantic convoy route. On 14 June 1941 the ninety-nine year lease on Argentia became official. Shortly after, Vice Admiral Arthur L. Bristol, Commander Task Force 24 (based in Argentia) arrived. He resided on and commanded his forces from his flagship, the USS *Prairie*. Captain G. Morgan, USN, was the first Commanding Officer. The Naval Air Station was commissioned on 28 August 1941. US Navy and US Army Air Corp

This Long Island Steamer was brought up from the Hudson River to act as a barracks for American civilian contractors and engineers. The Peck *arrived at Argentia on 18 January 1941. She had a capacity of 1500 personnel. The ship left Argentia in February 1943.*

This landmark is called Isaac's Head. During the war years, the American flyers warmly referred to it as "Mae West." This set of islands was part of the original leased land given to the US government by the British in the American Bases Agreement. This photo was taken in 1939 in the Virgin Point area of the Peninsula.

The original US Marine Detachment in 1941, standing inspection in foul weather. These marines landed at Argentia in January 1941.

planes were flying off asphalt runways at that time; all the runways were not completed until the following year.

The following narrative was written by a US naval historian stationed at Argentia:

> The Naval Operating Base and Naval Air Station was commissioned in 1941 and was thus in full operation when the United States went to war in December of that year. It was destined to play a significant part in the winning of that war. Throughout the entire struggle it was an Operational Headquarters for forces engaged in combatting the German submarine menace. For two years a Fleet Air Wing operated anti-submarine air patrols over the North Atlantic from Argentia. Crippling blows were struck against U-boat packs, and the famous "sighted sub, sank same" message was sent by a Navy pilot on patrol from this base. It was a Coast Guard manned squadron of Navy PBYS based at Argentia which flew patrols over Greenland and located the German weather stations set up on that coast. Throughout the war the drab, but vital Ice Patrol, by sea and air, warned of iceberg menaces to Naval Forces and convoys in the North Atlantic.
>
> Argentia was ideally located to serve as a base for supplying, repairing and servicing the fleet and Naval Forces of our Allies. Throughout the years after Pearl Harbor they came, hundreds of them, British, Norwegian, Panamanian, Danish, French, Dutch, Canadian, Russian, Greek and finally the Germans themselves, captured and in defeat, destroyers, submarines, service ships and personnel, with the Stars and Stripes flying in triumph over the Swastika on surrendered staffs, on their way to the United States, and internment. Almost every type of US Naval vessel steamed under "Mae West" and into this port, great battleships like the *Iowa, Alabama* and *South Dakota*, fleet aircraft carriers such as the *Ranger, Bogue, Card, Croatan, Mission Bay, Tripoli,* and *Core,* cruisers, destroyers, destroyer escorts, frigates, corvettes, patrol craft, Coast Guard cutters, seaplane tenders, repair ships, transports, reefer and cargo ships, fleet tankers: in short all of the varied units which go to make up the United States' Naval might.
>
> Activity reached a peak during the last spurt of the German underseas fleet in the spring of 1945. During the months of March and April of that year one complete task force and a dozen fleet and Allied task groups were supported by Argentia. Issues of provisions reached a record average of nearly 2,000 tons a month, the entire stock of diesel oil turned over completely within five days during one week of this period, an estimated 2,000 tons of general stores were issued, the ship's store cafeteria broke all records in numbers of persons served daily, and stocks of small stores and ship's store items were exhausted faster than they could be replenished.
>
> Among the many important units based at Argentia was the Atlantic Weather Patrol consisting of small vessels specifically equipped to record and report weather

On 17 September 1941, five US destroyers from Argentia met a merchant convoy and escorted it across the Atlantic. US protection for Allied shipping had officially begun.

The USS Abemarles, *a sea plane tender, arrived in Argentia 15 May 1941 to establish a base for patrol and support operations.*

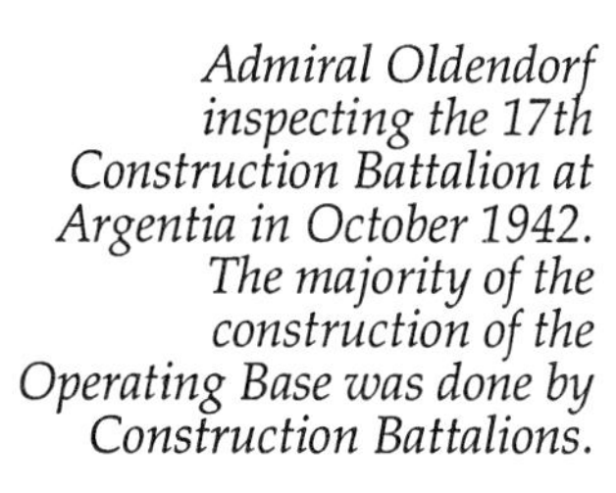

Admiral Oldendorf inspecting the 17th Construction Battalion at Argentia in October 1942. The majority of the construction of the Operating Base was done by Construction Battalions.

Some German vessels were captured and brought into the port of Argentia. Two captured German U-boats were brought into port with prize crews aboard.

Church service on board the HMS Prince of Wales, *10 August 1941, while anchored off Argentia Naval Base.*

observations. This force stationed ships at strategic points all over the North Atlantic and furnished ships and planes with late and accurate weather reports from one of the great weather breeding areas of the world. This service materially contributed to the safe and expeditious movement of troops and supplies across the ocean.

Following V-E Day an immense program of redeployment of forces began and movement of troops from Europe by air was heavy. The most travelled air lanes led right across the North Atlantic, and vessels based at Argentia were spread out all over the North Atlantic in air-sea rescue task groups to guard the sea under the air lanes and rescue personnel from planes forced down at sea.

Another story not be overlooked is the support rendered by this Naval Base to the US Army forces stationed in Newfoundland. Argentia is the only deep harbour in the country which remains ice free throughout the year. For this reason it necessarily is the point of discharge for cargo vessels and tankers, and the Navy had assumed the responsibility of getting fuel and provisions to Argentia for the Army. Throughout the war hundreds of railroad tank cars, box cars, and refrigerator cars moved from Argentia to the large Army posts and airports of the Trans Atlantic Ferry Command, supplying the all important logistic backing required to keep those vast and important units operating.

Use of the convoy system probably solved the problem of supply for European forces of the Allies and thus won victory over Germany. Argentia was a terminus for the scores of American and Allied warships on convoy escort duty. This was the western "Turn-Around" point for the Men-of-War, where they refuelled, took on stores, were "briefed" on tactics and operations, and formed their groups to take over the convoys coming across from the States. Thousands of tons of shipping were scheduled through submarine infested water from Newfoundland. One highranking British officer paid high tribute to the value of Argentia when he remarked, with sincere emotion, "Without Argentia, I doubt if we could have gotten over the hump that meant the winning of the war."

On 17 September 1941, five US destroyers from Argentia met a merchant convoy and escorted it across the Atlantic. US protection for Allied shipping had officially begun. Destroyers based in Argentia were among the first in the US Navy to have SONAR gear for the tracking and localization of submarines. On 14 December 1941 Germany and Italy declared war on the United States. The German U-boat threat was the most immediate threat to US mobilization.

October 1942 marked the arrival of the 17th Naval Construction Battalion, part of the 10th Construction Regiment, consisting of 1,049 personnel. They assumed construction responsibilities and base maintenance from the civilian construction companies. These sailors were commonly referred to as the "Seabees."

Submarine net, Little Placentia Harbour. This photo shows the area where President Franklin D. Roosevelt and Prime Minister Winston Churchill met and shared views on the world situation. The Atlantic Conference took place at moorings just off Virgin Point and between Argentia and the Issacs.

The US Navy aircraft carrier, the USS Mission Bay, *tied up to the wharf at Argentia. This wharf could accommodate large ocean-going ships.*

US Navy aircraft maintenance crew preparing the plane for patrol duty. This photo, taken during a blizzard, shows the crew manually turning the props prior to starting the engine. Many of this type of patrol aircraft were stationed at Argentia.

August 1941, on board the HMS Prince of Wales. *President Roosevelt and Prime Minister Winston Churchill discussed events involved in the Atlantic Conference. Standing behind and to the right of the President are Admiral King and General Marshall. Behind Prime Minister Churchill is Admiral Stark.*

US President Franklin D. Roosevelt stops to converse with crew members while involved in the Atlantic Conference in August1941. Both flagships were anchored in Placentia Bay with a backdrop of Issaac's Head. Both the Augusta *and the* Prince of Wales *can be seen in this picture, with a gangway between the ships.*

Argentia 1942-43, showing cold storage and supply buildings. The proximity of the main pier made it easy to off-load cargo and store it close by. The Marginal Wharf was popular with the Man-of-War ships that frequented Argentia, mainly for the job of escorting supply ships to England. The buildings at the top right were used for supplying these ships.

The Lighter-than-Air squadron operated from the US Naval Air Station at Argentia on anti-submarine patrol and ferry duty. This photo shows a US Navy blimp coming in over Isaac's Head. The photo below, 1944 vintage, shows the ground crew during the landing operation.

During 1944 the threat from German U-boats was somewhat lessened. The first of many German prisoners to be brought to Argentia by patrol ships arrived in November 1944. Spring of 1945 saw a final German submarine offensive, "Operation Wolfpack." In April an Argentia-based destroyer, the USS *Frederic C. Davis*, was torpedoed south of the Flemish Cap; 115 of 192 crew members lost their lives. A twelve-hour hunt ended with the capture of the responsible U-boat, *U-546*, and thirty-three prisoners.

The overall defence of North America, in the event of war, was further discussed in the United States and England between 4 and 18 August 1941 at the Atlantic Conference, in which the principles were placed on paper, but were never formally signed. These principles were called the Atlantic Charter, and later became the basis for the United Nations Charter. In view of these negotiations, both the President of the United States and the Prime Minister of England agreed to meet at a secret rendezvous in the Atlantic area, to discuss further the Atlantic Conference and share views on the world situation. The President of the United States, Franklin D. Roosevelt, arrived at Isaac's Point, just off Argentia, on 9 August 1941, accompanied by the battleship *Arkansas* and his flag ship, the USS *Augusta*. The same day, Sir Winston Churchill, Prime Minister of Great Britain, arrived on the HMS *Prince of Wales*. Special ceremonies were held aboard both ships on 10 August 1941.

"Sighted sub, sank same" was the message radioed to Argentia by Aviation Machinist Mate Donald F. Mason. He actually missed his mark, but his message was indicative of the mission and the mood of the patrol squadrons. Credit for the first U-boat sunk by Argentia forces was given to Ensign Tepuni in March 1942. In April 1942 Admiral Bristol died aboard the USS *Prairie*; Vice Admiral Brainard assumed CTF 24.

Expansion of facilities and commands continued in the years 1943 and 1944 in Argentia. A Royal Naval Air Station was established by the British; a $3 million ship repair unit and a 7,000 ton floating dry dock were completed; recreation facilities, a gymnasium and the North East Arm Recreation Camp were completed. Argentia served as base for the shakedown cruise for some of the Navy's newest battleships, the USS *Indiana*, the USS *South Dakota*, the USS *Alabama*, and the USS *Iowa*. That winter, a British anti-submarine task force of two carriers, was stationed in Argentia. At that time there were five US carriers and over fifty destroyers and patrol craft at Argentia. In the summer of 1944 a Lighter-than-Air Squadron was stationed in Argentia for ASW reconnaissance. In October 1944 an inventory and appraisal of the Argentia complex by local commands placed its value at $53,200,000.

Argentia was visited by many prominent individuals from government and from the field of entertainment during the years. In the late 1940s entertainment was provided by Phil Silvers and Frank Sinatra, among others. Government officials also made Argentia a

"Strain Hall," commonly referred to as the "Argentia Hilton," is the second tallest building in Newfoundland. The foundation was laid in 1955 and occupied in 1958. The facility, costing over $5 million, is the most modern of US Naval quarters for officers and enlisted men.

The main security gate at the entrance to Argentia Naval Station is manned 24 hours each day. Thousands of military and civilian personnel have passed through this gate since 1941.

stopping place. In 1953 thirty-three Senators and Representatives toured the facilities at Argentia as did Vice-President Nixon in 1958. Cardinal Spellman arrived on New Year's Eve 1958 and was greeted by typical Argentia weather; a four-day blizzard. Later that month Soviet Premier Anastas Mikoyan was forced to land in Argentia due to engine trouble on a commercial flight. The Christmas USO show in 1961 at Argentia included Jayne Mansfield and Bob Hope.

The Argentia Naval base today, although reduced in size considerably compared to the 1940-50 era, still remains as an active facility. It is the only US military installation left in Newfoundland. The mission of the base is primarily surveillance of submarines and other foreign naval vessels; the military attitude still exists, and continuously brings back great memories of fifty years ago. The towns around the base area which were very small in 1940 have grown to prosperous and modern communities. An occasional US sailor's uniform can be seen on the streets, reminding the middle-aged and the elderly of the "good old days" when the activity of Argentia was at its height.

Argentia was visited by many prominent individuals from government and the entertainment industry during the war years. In the late 1940s entertainment was provided by Phil Silvers and Frank Sinatra, among others.

The greatest problem encountered in construction, aside from the often inclement weather, was the removal of peat. Before work began, it was estimated that most of the land was covered with one to twelve feet of peat, with areas in which the peat was twenty feet deep. Final estimate was that 8,500,000 cubic feet of peat, gravel and earth were moved before construction of the runways was started.

Headquarters, Naval Operation Base, Argentia, 1943

3

Fort Pepperrell

On 15 September 1940, less than two weeks after the approval of the Lend-Lease Agreement, the USS *St. Lewis* arrived in St. John's with Rear-Admiral Greenslade, and other US military and civilian personnel, to confer with members of the Commission of Government. Rear-Admiral Greenslade was appointed as head of the United States-Canada Permanent Joint Board of Defence, which was set up by President Roosevelt on 18 August 1940 during negotiations of the Anglo-American Agreement. Some of the group accompanying him were technical experts from the US Army and Navy, who made a number of reconnaissance flights over Newfoundland to finalize the possible sites for the proposed US military installations. Upon conclusion of these meetings in St. John's, the Board of Defence members, including members designated within the Commission of Government, went to Canada for a Joint Defence Board conference. A group of US Corps of Engineers, under command of Lt. Col. Philip Burton, arrived at Argentia on 13 October 1940 aboard the USN *Bowditch*. Within a few days, Col. Burton and his engineers set up headquarters in the Reid Building on Duckworth Street, St. John's. They began the official surveys of all proposed sites and designed the layout, plans and specifications for construction. The Corps of Engineers supervised the construction at Fort Pepperrell, Fort McAndrew, Harmon Field and the American side of Gander and Goose Bay, Labrador.

An area close to St. John's was selected for many reasons: the seat of Commission Government was located there; the surrounding area was well secure with high cliffs on the ocean side; there was a large section of land available to meet the needs of a large military installation; an airport was close by for logistical support; and St. John's was the capital city of Britain's oldest colony. The "American Bases of Newfoundland Act," Newfoundland Statute of June 1941, described the total area, as previously surveyed and requested in the

This 1927 photo of Quidi Vidi Lake shows the area of Pleasantville, The Boulevard, and some of the farm buildings on what was to become Fort Pepperrell. Quidi Vidi Gut Bridge is in the foreground.

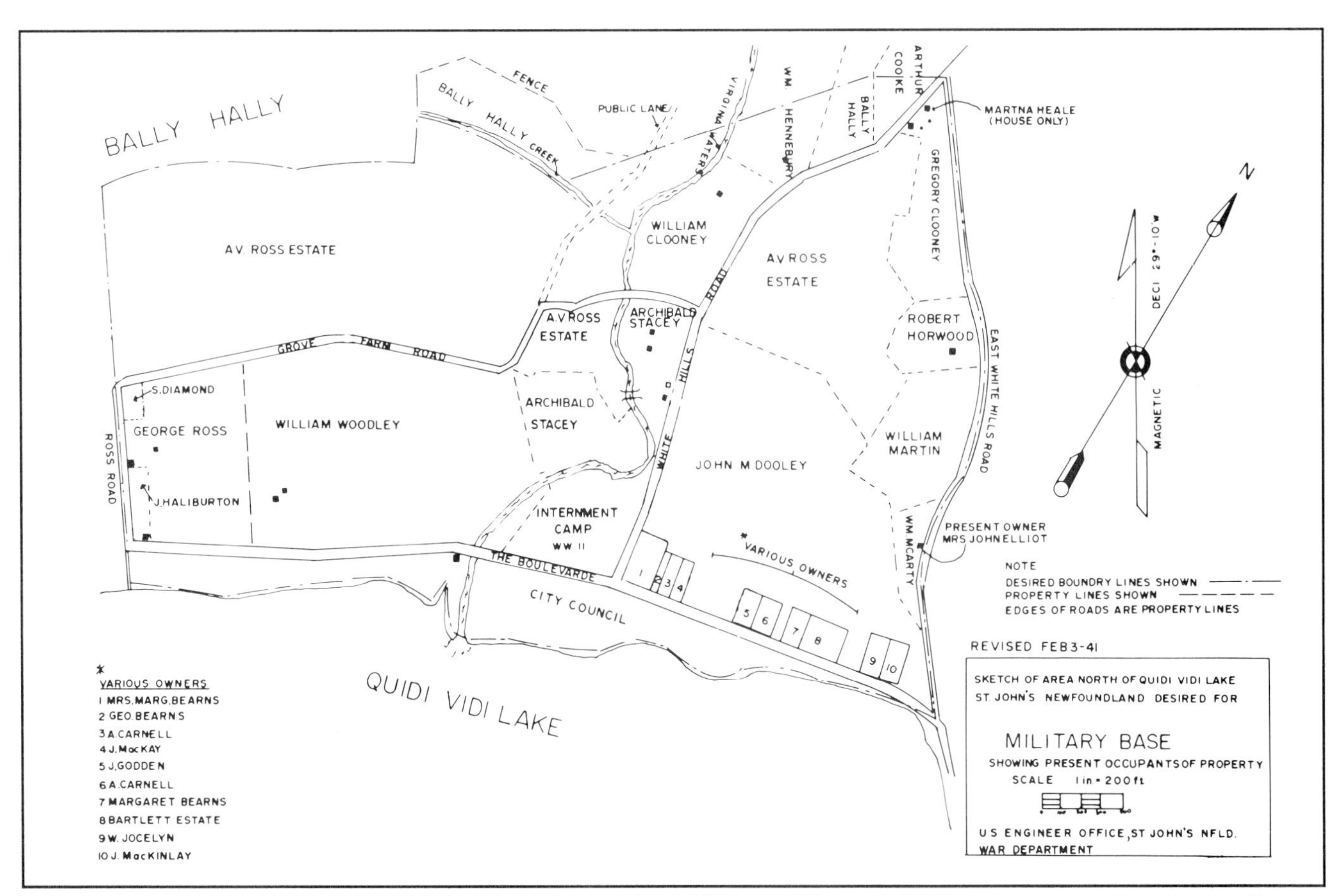

Map of property and property-owners around Quidi Vidi Lake, in the area of the proposed US Army Base, Fort Pepperrell.

Anglo-American Lend-Lease Agreement. The area consisted of 191.0 acres on the north side of Quidi Vidi, 7.0 acres which included Quidi Vidi Cove, .36 acres of Quidi Vidi lakeside, 2.44 acres known as the White Hills, and 22.2 acres of area known as Radio Tower Road. In addition, the Signal Hill Battery was included and involved 2.5 acres. Altogether, 227.5 acres of land in and around the St. John's area were designated for US military development.

The main area where Fort Pepperrell was to be built was, at the time, fully occupied with farms and several country homes on the north side of The Boulevard. Included as well was a new German World War II Internment Camp, located on the north side of Quidi Vidi, almost in the exact spot where the Royal Canadian Legion Branch #56 is located today. All of this land had to acquired. The land owners were all notified of what was to take place in view of the war effort and in defence of the Newfoundland area. A settlement was made in early 1941 by the United States Government to each land owner. They were advised to move out immediately. The twenty-five German sailors from the *Christoph V. Doornum* placed in the Internment Camp were supposed to have been moved to a newly constructed site in the Salmon Cove-Victoria area in 1941. However, it is unknown if this ever took place. It was rumoured they were transferred to another internment camp in Canada.

About mid-October 1940, the USN *Bowditch*, from Argentia, arrived at the Bursey Fish Plant premises on the south side of St. John's Harbour and aboard were a number of US military and civilian personnel from Argentia. They wanted to see Mr. W.J. Bursey regarding rental of his premises. The following is an extract from the book *The Undaunted Pioneer—An Autobiography by W.J. Bursey* (1977) Chapter 34, titled "The *Edmund B. Alexander*."

> ...The wind was from the northeast. It was cold and raining one afternoon in October of 1941 that a navy boat came to my dock bringing United States engineers who landed when my men were washing fish. The spokesman asked who owned the premises and where could they find him. I had an office somewhat back from the water. It was pointed out and he was told I was in.
>
> I was shocked to receive the visitor, dressed as he was in United States Officer's uniform, decorated with much gold braid. He introduced himself and told me that he wanted the premises for the United States Armed Forces, and that he wanted it by "tomorrow noon."
>
> They had been looking for deep water and had gone all around the Harbour with surveyors and that my place had the only water deep enough for their purpose.
>
> He told me that they had a large ship ready to leave for St. John's and asked how quickly I could move out. I took him on a tour of the premises. He saw more than 1,000 quintals of salt fish in bulk and a similar amount of pickled fish that was

ordered by the Gorten Pew Company of Gloucester Massachusetts. A United States Merchant Ship called at our wharf regularly to load for Boston. The pickled fish must be packed in casts and be ready when she calls.

I took him to the cod-liver-oil factory. We had there eight or ten thousand gallons of refined cod-liver-oil, some in tanks and some in casts. Besides there was the pressings in barrels and in vats and they were accumulating every day.

The Officer was impressed with the extent of our business and saw the impossibility of a quick move for us. We went back to my office and there he promised that they would cooperate with me in all reasonable ways. But he advised me to see my lawyer as soon as possible. Which I did. At that time Richard Cramm was doing my legal business. He drafted an agreement in legal terms as we had talked out in my office. We went together to the United States office that was then in the Reid Building on Duckworth Street. There we saw a drawing of the *Edmund B. Alexander* moored alongside my wharf and fastened fore and aft to rig bolts in the cliff of the Southside Hills. They read the legal agreement, approved it and we signed it there and then.

They would pay the agreed rent for the ship lying alongside my premises. They would give me the contract to supply the ship, their bases and camps and anything else under their charge.

They were indeed very cooperative. They had a line of dolfins for the ship to lie by. They used long wooden piles driven into the bottom outside our wharf and this gave us room to use our premises employing small boats for receiving and shipping our fish and oil.

When the ship arrived she brought one thousand troops from all over the United States. She was the largest ship to enter St. John's up to that time and can be remembered as the *Edmund B. Alexander*. She was piloted into St. John's by Captain George Anstey and brought alongside our premises. Captain Anstey had sailed three of the Hickman Company foreign going fleet. But he looked upon his piloting of the troop ship through the narrows where there were only two feet of water between her keel and the bottom as the most wonderful experience of his life.

Ours was a three storey house facing towards the waters of the Harbour and the heart of the city. But when the *Edmund B. Alexander* was tied up, the ship was about all that we could see. We couldn't see the water or the city even from our topmost windows. She remained at my dock for eight months.

The *Edmund B. Alexander*, once named the SS *America*, was built for the Hamburg American Line in 1905 by Barian & Wolf, of Belfast, Ireland. The vessel had a length of 669 feet, a beam of 74 feet and attained a speed of 17 1/2 knots. Her gross tonnage was 22,600. In August 1914, at the outbreak of the First World War, the ship was in Boston and remained there until the United States entered the conflict. The ship was seized and put into service as a US Transport

UST Edmund B. Alexander *upon arrival at St. John's, Newfoundland, 29 January 1941.*

under the name *America*. After the war, the ship was converted to passenger service for the US Mail Line and after a brief career, the company failed and the *America* went to the United States Lines.

In 1925 the vessel caught fire and was nearly destroyed, but was again refitted for trans-Atlantic passenger service. The US Government again acquired the vessel in 1931 and she was laid up in Chesapeake Bay until 1940 when a fourth conversion fitted her out as a troop transport. She was renamed the *Edmund B. Alexander*. Further work on the ship in 1942 gave her oil burning equipment and she continued in service during the war. The vessel was laid up in the Hudson River Anchorage of the National Defence Reserve Fleet as of 12 July 1955. Her present status is unknown.

It was called "The Friendly Invasion" when, early on the morning of 29 January 1941, the UST *Edmund B. Alexander* steamed through the Narrows of St. John's and moored up to the W.J. Bursey wharf premises on the Southside. Under command of Capt. W. Joensen, the ship had a crew of 200, and carried 977 US Army troops consisting of 58 officers and 919 enlisted men. The Commanding Officer of the US Army troops was Colonel Maurice B. Welty. The troops consisted of the US 3rd Infantry, 24th Coast Artillery and the 62nd Anti-Aircraft Artillery. They all came from the Second, Third and Seventh Army Corps of New York, New Jersey, Delaware, West Virginia, Pennsylvania and Maryland. The *Edmund B.* also carried 2,000 tons of cargo consisting of armament and general construction materials. The following is an extract from an article in *The Evening Telegram*, "Offbeat History," by Michael Harrington, 7 February 1983, on the arrival of the UST *Edmund B. Alexander*:

> The weather was very stormy and the visibility as bad on Saturday morning, 25 January 1941, that the ship could not be seen at Cape Spear....
>
> ...The Customs cutter *Shulamite*, in charge of Captain John Whelan, Inspector of Revenue Protection, with Captain Martin Dalton of the Newfoundland Railway and Captain Hiatt of the US Corps of Engineers, left the King's Wharf and headed out to sea. Also aboard the cutter was Master Pilot Captain George Anstey and press representatives from *The Evening Telegram* and *Daily News*.
>
> As the *Shulamite* cleared the heads, the first glimpse of the troopship was seen through the snowdrift, about a mile offshore....
>
> ...The troopship presented an impressive sight in the snowy daylight. Painted grey, her hull seemed all the greater and in spite of the considerable swell she was immovable. Her five decks, topped by four masts and two tall funnels, gave an impression of incredible height. The funnels were painted a buff colour with bands of red, white and blue, topped off with a band of black paint.
>
> ...The troopship headed for Conception Bay to seek shelter from the elements in anchorage between Bell Island and Kelly's Island. She remained there over the weekend until Tuesday when the weather cleared and she again rounded Cape St. Francis and approached the port.
>
> Many citizens had assembled at various vantage points to see the big ship coming in; the *Edmund B. Alexander* was larger than any ship that had entered St. John's up to that time.... The American Troopship was 668 feet long, with a beam of 74.3 feet and a depth of 47.8 feet and displaced 21,000 tons....

Mr. Herb Wells, in his "Vet's Column" in *The Evening Telegram*, wrote the following regarding the reaction of the general public of St. John's the day the *Edmund B.* arrived.

> ...Because of the delay, the population of St. John's became aware that the Americans were soon to land and preparations were made to receive them with some degree of hospitality. Mercantile houses along the harbour front displayed bunting, and some homes throughout the city were decorated. This gesture may have seemed to some to be unusual—decorated buildings to welcome at wartime. Newfoundland, as part of the British Empire, was at war with the Axis Powers, America wasn't.

Dance aboard the Edmund B. Alexander *in 1941. Many of the local people were invited guests.*

US soldiers board for Newfoundland at New York. An officer gives last minute inspection of rifles of troops bound for Newfoundland on the Army transport Edmund B. Alexander, *which left the Brooklyn Army Base 15 January 1941.*

US troops board the Edmund B. Alexander, *formerly the German liner SS* Amerika, *seized during World War I. The voyage to Newfoundland took five days.*

The port side of the Edmund B. Alexander, *docked in St. John's Harbour on the Southside, showing several US soldiers going into town on passes. Many of the local fishermen of St. John's used their small fishing boats to operate a ferry service between the north and south sides of the harbour.*

After an elaborate refitting of the ship at Baltimore, Maryland in 1940, she was renamed after Commander Edmund B. Alexander, who commanded the 3rd US Infantry in the Mexican War of 1846-1848. The ship had all the conveniences of a passenger liner. There were individual rooms for officers and double room accommodations for enlisted men, spacious dining halls for officers and soldiers, individual recreational rooms for each designated organization, cafeterias, a well-fitted hospital and dental area, a swimming pool, a large auditorium for movies and joint briefing sessions, offices, a library and reading rooms, a pool room, bars and many other facilities.

The US Army soldiers and many of the officers used the *Edmund B.* as a floating barracks from the time of arrival until July 1941. Construction was just beginning on Fort Pepperrell, which was to be their military garrison upon completion. The Commanding Officer, Col. Maurice B. Welty, established his Command Post at 44 Rennies Mill Road. Several of the high-ranking officers resided in homes and other rented places throughout St. John's. The UST *Edmund B. Alexander* left St. John's in July 1941. She returned again in 1947 with more troops and military supplies for Newfoundland.

In March/April 1941 work began on the construction of a temporary camp for the military personnel aboard the *Edmund B.* An area between Rennies Mill Road and Carpasian Road was selected, called "Camp Alexander," and had to be large enough to temporarily accommodate not only the number of troops on the *Edmund B.*, but those expected into St. John's on other ships. One of these was the UST *Leonard T. Wood*, which arrived in May 1941. On 15 April 1941 a lease was signed with Carpasian Park Limited for 15 acres of land on which to build Camp Alexander. On 20 May 1941 the troops moved into their new but temporary quarters. Some of the Coast Artillery personnel moved to their Signal Hill accommodations. They remained there until 14 December 1941 when the first permanent barracks and other vital facilities were ready at Fort Pepperrell.

Two members of the US 3rd Infantry, who arrived on the *Edmund B.*, were quartered at Camp Alexander, and currently live in St. John's with their families, recall their stay there. Mr. Russell Englund and Mr. Cecil Hutchens trained hard on the now St. Pat's Ball Park area, and bivouacked in many areas in and around St. John's. Life at the "Tent City" wasn't bad at all during that summer and fall; besides frequent three-day passes in the St. John's area, everyone had access to the recreational facilities at Camp Alexander. Guard duties at strategic locations and 24-hour watch by troops at the many observation tower posts in and around the vicinity of St. John's were the order of the day for most troops. Some Army troops were transferred to the Argentia area in mid-1941 for security duty

Photo shows erected tents at Pleasantville in 1914, where the Royal Newfoundland Regiment trained before departing for World War I.

The Officers' Club at Camp Alexander. Left to right: Capt. John Moorehead, Lt. Wm. Fleming, Capt. Jos. Curiello, Lt. Irving Held, Capt. J.S. Kajawski, Lt. "Gumps" Taylor, Lt. T. Hansen, Lt. "Skeets" Sexton and Lt. P. Rose.

Lines of tents at Camp Alexander showing dirt road and boardwalk in front of tents, 1941.

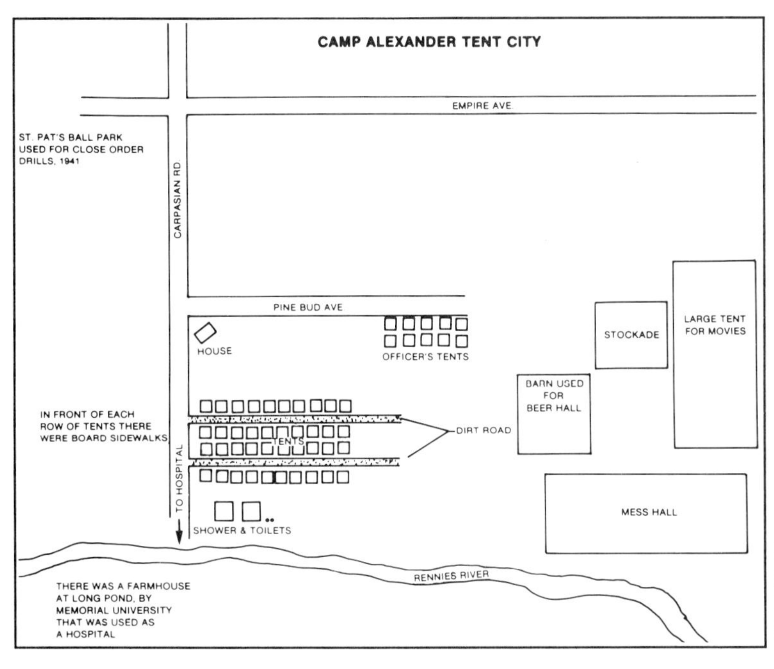

Camp Alexander tent city. The US Army temporary military hospital was located at Long Pond, north of Camp Alexander, in the former home of the Carney Family.

there and to supervise the erection of the big guns that protected the shoreline.

Immediately after the arrival of the *Edmund B.* many of the US Coast Artillery and Anti-Aircraft Artillery personnel were engaged with local contractors in the construction of temporary barracks and other necessary buildings on Signal Hill. The first barracks, mess hall and other vital support buildings were ready to be occupied by May of 1941. The positioning of anti-aircraft and other large guns on Signal Hill, and at other areas in the immediate locality, was begun as soon as such materials were unloaded off the *Edmund B.*, or as soon as they arrived on other supply ships.

Fort Pepperrell was named after Sir William Pepperrell, who was born at Kittery Point, Maine in 1696, the son of an English fisherman who later became a great shipowner and merchant. Sir William greatly expanded his father's thriving business and, at the age of 26, he was a colonial leader—a Colonel in the Maine Militia. In 1745 he led one of the boldest attacks in colonial history when, with only 100 small vessels, he sailed under the very noses of the French into Louisburg harbour, effected a landing, besieged the fortress, and forced the French to surrender. For this outstanding service he was created a baronet, Pepperrell of Massachusetts, the following year. He died in Kittery on 18 July, 1759.

The original group of Newfoundlanders and Americans who worked for the Corps of Engineers during the construction of Fort Pepperrell. Some members of the press are also in this 1942 photograph. Over 300 worked for the US Corps of Engineers.

A 1942 military parade with the US Army 3rd Infantry taking the salute as they pass the grandstand by Government House on Military Road. Governor Walwyn can be seen on the stand. The history of the 3rd Infantry dates back to the Revolutionary War. Their colour guard is authorized to use uniforms of that period.

The general public was astonished when, during a joint American-Canadian military parade in 1942, these tanks came into view. The US Army had twelve of these tanks assigned to the military at Fort Pepperrell.

Construction at Fort Pepperrell (APO 862) began on 8 February 1941 under the direction and supervision of the Newfoundland Base Contractors (American Company), and the US Corps of Engineers. The first American general contractor was Al Johnston Construction Limited from Hibbing, Minnesota, USA. Hundreds of Newfoundlanders were employed. Engineers, draftsmen, carpenters, plumbers, electricians, labourers, heavy equipment operators, other vehicle drivers, office workers and others made up the civilian work force. The overnight growth of the city of St. John's and the economic boom was phenomenal. The effect was felt all over Newfoundland as men and women left their communities for St. John's in search of work at Pepperrell. By the middle of 1942, over 2,200 civilians were employed with over 400 American civilians. Over 5,500 civilians were employed at the peak construction period.

Plans called for the site to accommodate 3,500 troops, inside storage space for 310 vehicles, warehouses totalling 14,315 square feet, and 20 acres of outside storage of materials. A large hospital, bakery, commissary, power plant, cold storage, laundry, fire station, guard house, maintenance shops, chapel, and armament repair shops were but a few of the facilities required. King's Bridge Road and The Boulevard were also upgraded in 1942.

In view of the escalation of the World War in Europe and the increased bombings of England by the Germans, the War Department in the United States set a time limit of December 1941 for the completion of the barracks and other support facilities for the military. Building materials had to be transported to Newfoundland by ship, and thence to the construction site. Argentia was also under construction at the time and its problems of transportation of materials and equipment were similar. The port of Argentia was blocked with vessels, so all cargo destined for Fort Pepperrell had to be brought into the St. John's port by sea or rail. Maximum use of the Newfoundland Railway was also employed, as much cargo came across the ferry at Port aux Basques and was unloaded in St. John's. The use of the Newfoundland Railway became limited in the latter part of 1941, when Stephenville Air Base was under construction. The US military decided it would be most advantageous to have its own docking facilities in St. John's, therefore, the project got under way early in 1941.

Fort Pepperrell was constructed on the natural topography of the area. Unlike Argentia, there were no large areas of peat bog to be removed. The area on the north shore of Quidi Vidi was rather level. This area, known at the time as "Pleasantville," was where the Royal Newfoundland Regiment trained before departing for World War I. A large number of temporary buildings were erected to serve as offices, warehouses, shops and other facilities necessary for the large construction project. One of the first buildings was a large wooden structure on the edge of The Boulevard by Virginia Waters, which was used by the Corps of Engineers and Newfoundland Base Contractors as a general office. This structure was later renovated and a larger section added to it. It was then used as the Pepperrell Recreational Building and contained a bowling alley, swimming pool, ping-pong and billiards rooms and a large upstairs area used as a joint service club. Today, this structure still remains, and is used as the Headquarters for Provincial Command, Royal Canadian Legion, and the Royal Canadian Legion Branch Number 56.

During construction, each road or street was alphabetically lettered: i.e. "A" Street, "B" Street, etc. Furthermore, each building was given a specific number, although these numbers were changed, as well as the street names, prior to occupancy.

Fort Pepperrell, although not completed, was ready to accept troops by November 1941. The 21st Signal Service Company and Companies "I" and "L" of the 3rd Infantry were the first units to move in. Camp Alexander was empty by 14 December 1941. Headquarters, Newfoundland Base Command, moved from its quarters at 44 Rennies Mill Road to Pepperrell on 19 February 1942. The first families of some officers and enlisted men arrived in St. John's in the early spring and summer of 1941. They found rented accommodations in and around St. John's, and many of them moved into new married men's quarters at Pepperrell in November and early December 1941. After the 7 December 1941 attack on Pearl

View looking northwest at one of the barrack buildings. The foundations for the fire station and guard house are shown in the right centre. The barrack-type building on the far left was to become the Headquarters building for the Newfoundland Base Command.

Harbor by the Japanese, an immediate order was issued by the US War Department for all American military dependents, in every location in Newfoundland, to return to the United States. They returned by train, ship and aircraft within ten days.

The building program for Fort Pepperrell was changed considerably after 7 December 1941. All temporary buildings for the construction forces were ordered left standing and a large part of the permanent construction was postponed. The capacity of troops was to increase to 5,500 or more. Those who were not quartered in the new permanent barracks were placed in modified temporary buildings belonging to the contractor. The new barrack-type buildings could comfortably accommodate 125 men in single bunks. There were many occasions when the single bunks were changed to double units, with one on top of the other. Influxes of military personnel on many occasions required the erection and re-use of some of the tents used at Camp Alexander to accommodate them. These tents were set up on the spacious parade ground located to the rear of the chapel building. US military records indicate that in 1941-43 there were as many as 7,000 troops at one time or another at Fort Pepperrell.

On 14 December 1941 the former private yacht *North Gaspé* moved out of Brooklyn, New York Port of Debarkation and arrived in St. John's on 19 December. She carried fifty Weather and Army Air Communication technicians and Federal Bureau of Investigation agents. Nineteen of the AACS men were immediately formed into

This 1944 view from the boathouse area on Quidi Vidi shows the well-known horse and sleigh races on Quidi Vidi Lake in winter. Note the west end of Fort Pepperrell showing part of the West Motor Court, Base Theatre and some temporary contractor buildings still erected on the site.

the 8th Airway Communication Squadron and were assigned to the four major stations in Newfoundland, namely Harmon, Torbay, Goose Bay and Gander. The remainder of the military technical personnel were part of the 685th Air Warning Squadron, who later set up five top-secret US Radar Stations in Newfoundland. The Navy at Argentia had its own radar personnel assigned.

As more facilities became available at Fort Pepperrell and other US military locations throughout Newfoundland, additional troop ships continued to arrive at St. John's and Argentia. In April 1942 the UST *Shadow Theory*, sailing from New York, arrived in St. John's Harbour with 200 sailors and 1,800 soldiers. The *Edmund B. Alexander* made a second trip and arrived in June 1947 with more US Army and US Army Air Force personnel. She continued on to Iceland with additional US military personnel.

The number of cargo vessels entering St. John's port during 1941, carrying thousands of tons of cargo for the Fort Pepperrell base construction, created a problem of space. Along with the property acquired for the US military in the St. John's area, the US Army received docking space at the Lower Battery, just inside the entrance to St. John's Harbour. There, in 1942-43, a magnificent pier and storage sheds were erected. The main shed is over 30,000 square feet. The wharf is 605 feet long by 160 feet wide and has a depth of water alongside of 31 feet. The US Army Dock was a very busy place during all the years Pepperrell was in existence. The overall cost of the Army Dock was $1,726,274.

Gasoline and oil for use by the US military were brought in by US Naval and other tankers and off-loaded into earth-covered (underground) storage tanks. Gasoline and oil products were transported by oil tank trucks daily from the dock to Pepperrell. They were a common sight each day on the St. John's east end streets, with a fire truck following the tank trucks. To eliminate this hazard, in 1957 a pipe system was installed running from the dock up over the Battery and Signal Hill, down to the new storage tanks at Pepperrell, at the foot of the Quidi Vidi Lake area. Furthermore, an under-wharf piping system for gasoline and oil was installed. The products were pumped from the tanker ship to the storage tanks at the dock and then, as required, pumped through the supply line at Pepperrell. This eliminated the use of the mobile gas and oil tankers on the streets.

While Fort Pepperrell was being constructed in 1941, great activity was first observed at Signal Hill. The US Coast Artillery, the Anti-Aircraft Artillery, and contracting personnel were busy building a military outpost on the Signal Hill area. The site was where the National Historic Park Museum building is today. The facilities which were independent from Fort Pepperrell included accommodations, mess hall, recreational building and other structures necessary for the troops to carry out a twenty-four hour alert, to maintain and operate the eight and six-inch guns installed along the perimeter of Signal Hill. There were as many as 1,000 or more troops stationed there.

The following is an extract of information from "A Structural and Narrative History of Signal Hill National Historic Park and Area to 1945," by James E. Candow.

> Immediately after the *Edmund B. Alexander* docked, the US Army unloaded one four gun 155-mm mobile anti-aircraft battery, one four gun 3-inch anti-aircraft battery and one 16 gun battery of .50 calibre machine guns. These were placed in temporary positions on Signal Hill. The Americans added two 8-inch coast defence guns to Signal Hill in September 1941. It is assumed that one of the 8-inch guns was placed near Cabot Tower, in the area of the present parking lot. The guns on Signal Hill were manned by the 53rd Battery, United States Coast Artillery. 8-inch guns were also placed at Red Cliff Head, between Torbay and Logy Bay area, at Fort Cape Spear. (Other guns were placed at Flatrock and Robin Hood Bay.) The US Army also built an aircraft observation station on Signal Hill and manned this post on 1 February 1942. Friendly aircraft were required to approach Signal Hill station from the North on a course 270 degrees magnetic, at an altitude of 1,000 feet.
>
> St. John's escaped a major test of its defence during the Second World War. The only scare arrived on the afternoon of 3 March 1942 when a German submarine off the Narrows fired two torpedoes, believed to be for an ammunition ship anchored in the harbour. The first exploded at 2:54 p.m. in front of Fort Amherst, and the second went off just two minutes later on the opposite side of the Narrows. A general alarm was sounded at Fort Amherst and the guard doubled at Fort Chain Rock, but there were no more further developments. The submarine menace in Newfoundland waters was at its peak that year. U-Boats sank two ships anchored off Bell Island and two more in convoy there, but the worst disaster was the sinking of the Newfoundland-Cape Breton ferry *Caribou*, in which 136 people were killed. The U-Boats penetrated the Gulf of St. Lawrence and the Strait of Belle Isle in 1942, and 21 ships were lost there. [The area was nicknamed "Torpedo Junction."]

One of the many big guns protecting St. John's Harbour

Hundreds of US Army soldier hours were expended in training and improving camouflaged positions. From the period of 1941 to 1944, extensive training manoeuvres were carried out on the White Hills area, which today is known as Virginia Park, the White Hills and Lundrigan's Concrete Limited. Several ammunition igloos were erected on the site and throughout the White Hills area, as well as rifle and pistol shooting ranges. One old tank was used extensively for target practice. This training went on until 1944 as new units continued to arrive at Fort Pepperrell from the US. The twelve

Site of the proposed US Army Dock at the Lower Battery on 1 October 1941. Work started in the spring of 1941 and the facility was in use in 1942. Newfoundland Base Contractors were awarded the contract for construction on 8 February 1941. However, the Dock was not completely finished until early 1943.

United States Army Transport ship Fairfax *being moored at the US Army Dock, St. John's in August 1945. This was the first ship to bring the American soldiers back to the United States.*

This aerial view shows the US Coast Artillery and Anti-Aircraft Artillery accommodation units on Signal Hill. Note the tents used to house some of the personnel when the barracks were full. An ammunition Quanset Hut can be seen in the upper left corner.

One of the twenty-four barrack-type buildings constructed. These buildings were the most modern accommodations for military personnel ever built outside the US. They were self-contained, with large recreation rooms, mess hall, barber shop, tailor shop, PX, offices and rifle and pistol ranges.

The Base Chapel where worship for all denominations was held. The church records for 1960 indicated that over 2,000 marriages had been performed there since 1941.

The Base hospital, now The Janeway Child Health Centre. Constructed in 1943, the underground area was made gas and splinter proof so that all medical facilities could be moved below ground if necessary. The first two babies born at the hospital to Newfoundland mothers were Peter Cardoulis of St. John's and Charlie Cramm of Bonavista.

The USO building on Merrymeeting Road as it looked before the official opening in 1943. Today the building is being used by the Government of Newfoundland and private enterprise.

ammunition igloos constructed on the White Hills area were earth-covered, eighteen-inch concrete structures. Each was so carefully camouflaged with natural cover that it was impossible to detect its location from an aircraft. Most all of these are still intact today and some are being used by private enterprise.

The mission of the Newfoundland Base Command, as part of the Eastern Defence Command from 1941 to 1945, was to coordinate with the Canadian Armed Forces and secure the Island of Newfoundland from an enemy attack. From May 1945 to December 1946 the mission of the Newfoundland Base Command was threefold: to provide local security for and maintain all US Army installations and areas in Newfoundland; to facilitate operations of the Air Transport Command; and to maintain liaison with the US Navy at Argentia, allied nations and local commands. Effective 1 January 1946, the Newfoundland, Greenland, Iceland, and Bermuda Base Commands were relieved from assignment to the Eastern Defence Command and assigned to the US Army Air Corps under jurisdiction of the Atlantic Division of the Military Air Transport Service (MATS).

An old US Army tank used for target practice.

In December 1946 the prime mission of the Newfoundland Base Command included:

(1) the protection of national policy;

(2) the exercise of authority over all personnel, bases, installations and facilities under the control of the Atlantic Division and US War Department within the Newfoundland area;

(3) the coordination and supervision of air bases and related facilities included within national policy and strategic planning in the area;

Photos of some of the military training exercises at the White Hills area at Fort Pepperrell. All of the area–today known as Virginia Park, Harding Road and Lundrigan's Concrete (East) Ltd.–was used as a training ground.

Below, a group photo of 24th Coast Artillery, Battery "D," at Red Cliff in 1942.

(4) the consolidation, reorganization and reduction of all military activities in the area in accordance with demobilization plans and directives issued by the US War Department and base headquarters;

(5) the preparation and maintenance of plans for military defence of the base command area within the present resource and the coordination of such plans with local Navy and Allied Commanders; and

(6) assistance to the representatives of the Foreign Liquidation Commission in the disposal of surplus property within the area.

The command was further responsible for dispatching, servicing, supplying and maintaining all aircraft under operational control of the Atlantic Division scheduled or routed through or within the Command and had responsibility for the operation and disciplinary control of such air units, personnel and crew while within the command. The Commanding General was also directed to establish and maintain a separate search and rescue system.

On 1 January 1946 Fort Pepperrell officially became Pepperrell Air Force Base. In October 1947, under the US Congressional Unification Act, all the activities under the Army Air Corps were transferred to the newly-formed United States Air Force, and by 1949 all USAF personnel were wearing the new Air Force blue uniforms.

Fort Pepperrell in 1947 showing some of the temporary construction buildings in place. The structure shown centre left is the Base theatre. The building in the lower left is the fire station/guard house.

The Northeast Air Command (NEAC) was activated on 1 October 1950, and all units of the Air Transport Command stationed in Newfoundland, Labrador and Greenland, were transferred by General Order from the United States Joint Chiefs of Staff. This order also applied to all units and organizations, including civilian workers who were assigned to any remote station or area. The mission of the Northeast Air Command was to integrate more fully administrative and operational control of all US forces in Canada and Greenland. Further, the mission was to maintain and operate air bases, communications and weather facilities, navigational aids and an air rescue service to support the Strategic Air Command, Military Air Transport Service, and in coordination with Canadian and Danish Forces, defend these installations against attack.

Torbay Airport, officially opened by the RCAF on 15 December 1941, was jointly used by the RCAF, RAF, and the United States Army Air Corps until December 1946. Although the airfield was not used as much as Argentia, Gander, Stephenville and Goose Airports in the movement of large numbers of aircraft to England, it was still busy. The Royal Air Force had its own squadron of fighters, surveillance and weather aircraft stationed there. The RCAF personnel strength on the station during the peak war years was well over 2,000. Through an agreement between the US and Canadian governments

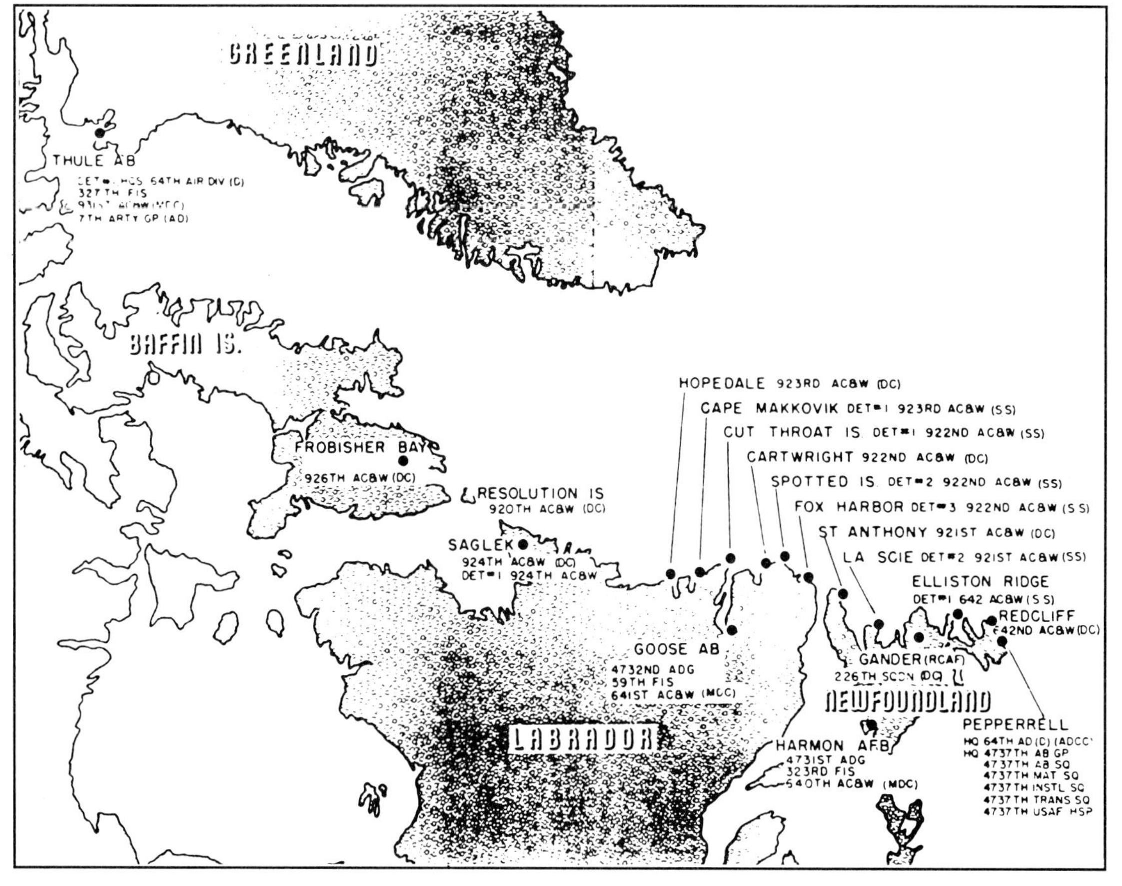

Geographical Location of 64th Air Division (D) Units.

early in 1947, the United States Air Force took over the use of the airport facilities and utilized about ten of the buildings located there. The US Military Air Transport Service (MATS) needed Torbay Airport in order to complete its assigned mission at that time. Maintenance of the airport and facilities was done by the Canadian Department of Transport. On 15 April 1953 the RCAF Station at Torbay was reactivated and RCAF personnel started to move in and began providing the necessary administration and operation of the facility to support the mission of its co-tenant, the United States Air Force. Early in 1954 a rental agreement was signed between the USAF and the RCAF and the USAF acquired the use of additional buildings. The 6600th Operations Squadron of the USAF was activated on 1 July 1953 and moved into existing Torbay Airport facilities. The USAF rented over thirty buildings at a cost of over $210,000 per year. The USAF operations at Torbay Airport from 1953 to 1958, when the Northeast Air Command was deactivated, were notable achievements. Thousands of tons of air cargo were handled by Torbay in support of the many bases and satellite stations throughout the Command.

On 1 February 1949, for the first time since World War Two, a ship was assigned by the New York Port of Debarkation to make scheduled runs between New York and St. John's. The ship was the 3,700 gross ton US Army Transport *Sergeant Jonah E. Kelly*. The *Kelly* departed St. John's on 7 February 1949 with her first cargo of

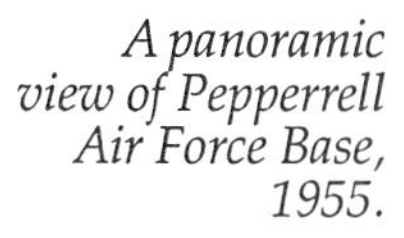

A panoramic view of Pepperrell Air Force Base, 1955.

US Army Transport Sergeant Jonah E. Kelly *anchored at the Army Dock, loading supplies for northern bases and sites.*

materials for northern bases and sites. She continued these trips up to 1961, when Pepperrell and the Army Dock were inactivated.

In August 1956 the Northeast Air Defence Command was abolished and on 1 April 1957 the Northeast Air Command was inactivated. Pepperrell Air Force Base then became the headquarters for the Air Defence Command's 64th Air Division, as well as all in-place organizational units of NEAC, the Transportation Terminal Command (Arctic) of the US Army, the Military Sea Transportation Service and the North Atlantic Army Air Communication Service (AACS) Region. The primary mission of the 64th Air Division was to the North American Air Defence Command (NORAD), to the Continental Air Defence Command (CADAC), and to the United States Air Defence Command (ADC).

The division's responsibility to NORAD was an operational one in that the work of actually guarding the North American continent comes under the jurisdiction of NORAD. The division accomplished this responsibility through the Northern NORAD region of Canadian operational headquarters at St. Hubert, Quebec.

The New York Air National Guard unit, the 152nd Aircraft Control and Warning Group, was called to active duty at White Plains, New York on 19 August 1951. In March 1952 it was designated the 64th Air Division (Defence), and advance parties were immediately sent to the northeast where they began to set up Aircraft Control and Early Warning (AC&W) Stations. On 3 November 1952, the first F94B jets of the 59th Fighter Interceptor Squadron arrived at Goose Bay Air Base, Labrador. Subsequent units were also sent to Ernest Harmon Air Force Base, Stephenville and to USAF Air Base

Top right: Armed Forces Radio Service, Station VOUS, Pepperrell, 1945. Most programs were recorded on record discs. However, two "super-pro" shortwave receivers were installed to receive programs broadcast directly from the United States.

Centre right: A private party, Communications Squadron, July 1954.

Bottom right: Scene from a Pepperrell NCO Club dance, 28 December 1949.

Below: Charlie McCarthy and Edgar Bergen were the unanimous choice of the soldiers as the top entertainment brought to the Island by the USO. They toured the Island in 1944 and visited practically every US military outpost. This photo shows Charlie being introduced by Bergen to the children of St. John's and some of the people of Quidi Vidi Village, where this photo was taken. A young boy shows them where they dry the fish and Charlie registers amazement.

at Thule, Greenland. All operations were controlled out of Pepperrell Air Force Base.

During the war years Newfoundland was visited by many famous movie stars who entertained the troops at each major installation. Bob Hope, Joan Blondell, Jayne Mansfield, Frank Sinatra, Frances Langford, Edgar Bergen and Charlie McCarthy, Andy Devine, Phil Silvers, Marlene Dietrich, Paulette Goddard, Victor Mature, Jimmy Durante and Rosemary Clooney were only a few of those who performed at Argentia, Gander, Pepperrell, Goose Bay and Stephenville. The United Service Organization (USO) originated in 1942 and was created to entertain American troops overseas. The USO constructed many buildings in the area of military installations, to provide a variety of entertainment for the US soldiers when they were outside their base areas. The first one in Newfoundland to be constructed was the USO Building on the corner of Merrymeeting Road and Bonaventure Avenue. It was officially opened in 1943. The facility had a separate area where the servicemen could sleep overnight, a large cafeteria, a spacious auditorium which was used for either roller skating or dances, and a games room and library. Many of the USO movie stars entertained the US personnel and their escorts here. Many Newfoundland girls served as hostesses in the USO building, before it was sold in 1946.

Over the years Pepperrell and the bases were host to many dignitaries. Her Majesty Queen Elizabeth II and Prince Philip visited St. John's in 1959 and spent some time touring the base. Other

Her Majesty Queen Elizabeth II and Prince Philip inspecting a US Military Guard of Honour on Water Street, St. John's in 1959.

distinguished guests included Cardinal Spellman, Edward G. Robinson, and Ed Sullivan. Many will still remember the Armed Forces Day activities held during May of each year. Open house was held on the base with various displays, parades, and parachute jumps into Quidi Vidi Lake. Special simulated aircraft crash fires at Torbay Airport were also included. One year, the base recorded over 6,000 visitors. Each base held an Armed Forces Day, with open house for the general public. Argentia included tours of large Naval ships at their dock. On most occasions, a US Naval destroyer would be tied up on the north side of St. John's Harbour, and the captain arranged for people to be conducted on a tour of the ship.

Toward the latter part of 1945, at the end of hostilities, the Newfoundland Base Command received permission from Washington to use two vacant buildings at Pepperrell as an NCO club from 1945 to 1947, at which time a barracks building was converted into a permanent NCO club. The Officers' Club was the former Officers' permanent transient quarters. It was opened in 1946. The enlisted men continued to use the recreational hall service club until the Enlisted Men's Club was opened later in 1947.

Many people from the St. John's, Stephenville, Argentia and Goose Bay areas remember the enjoyable times at the NCO clubs and the Petty Officers' Mess at Argentia. They will particularly remember the slot machines, foot-long steaks, 15¢ beer, 25¢ mixed drinks, 10¢ cigarettes, and 5¢ Coca Cola. The prices were cut in half on "Poor Richard" nights, and many a "40 ouncer" found its way off the base.

Radio Station VOUS (Voice of the United States) officially went on the air on 1 November 1943. The station operated on a moderately high frequency and covered not only the Argentia Naval Station and Fort McAndrew, but other US installations in Whitbourne, Colinet, Signal Hill and military observation towers along the East Coast. Many of the people who tuned in to this radio station will recall the opening and closing of each program: "This is the Armed Forces Radio Station." Also remembered would be the variety of radio programs such as "Suspense," "Command Performance," "Dinah Shore," and "Bing Crosby Hour." Other US military radio stations were installed at Harmon (Stephenville) and Goose Bay, Labrador. The local radio station at Gander was installed in 1944 by the RCAF Towards the latter part of 1960, the radio station VOUS transferred to Argentia, where it operates today on a very low frequency. Between each program is heard, "This is the voice of the American Armed Forces Information and Education Service."

Sports activities among the various military units, as well as with local sports organizations were active over the years. The US Army and Air Force sponsored many sports activities such as minor baseball and softball leagues. Youth teams were trained in football, roller skating, bowling, basketball and other sports. The most

Company "M" 3rd Infantry, Fort Pepperrell, winners of the American Servicemen's Race in the 1941 St. John's Regatta. Left to right: August Seacotte, Arthur Welsand, Arthur Groves, coxswain, John Cox and William Brown.

1941 Regatta.

Winning crew of the 1942 Regatta from "B" Battery CAAA. Left to right: Francis Yonkin, Col. Gaspare Blunda, John Fuller, Shotty Rogers, coxswain, Dennis Kelly, Capt. Ross, Battery Comdr. and John Bacha.

Lt. Col. Starling of Battery "B" accepting the Lady Walwyn trophy in the 1942 Regatta.

important sports activities were participation in the annual Regatta races on Quidi Vidi Lake. The large parade grounds at Pepperrell were used for many sports activities and today are used primarily for that purpose. One of the pioneers of sport competition between the US military and local units was T/Sgt. Charlie Riddle. Charlie arrived here in Newfoundland on the second voyage of the UST *Edmund B. Alexander* in 1947, married a Newfoundland girl, and has made his home in St. John's ever since.

It was indeed a sad day for the populace of St. John's and the surrounding area when, on 11 August 1961, Pepperrell Air Force Base officially closed. The following is an excerpt from the *Daily News* dated 12 August 1961 as written by Eric A. Seymour:

> Pepperrell Air Force Base faded into history yesterday afternoon following a brief flag ceremony. There was no speech-making. The US Naval Band from Argentia contributed the most important and touching aspect of the ceremony playing beautifully *My Maryland, O Canada, The Queen,* and *Ode to Newfoundland.*
>
> It seemed significant to us that the final anthem as the flags, the Union Jack for Newfoundland and the Canadian Red Ensign, for Federal Canada, reached the top of the mast, as the *Ode to Newfoundland* was being played!
>
> There was a guard of honour from Red Cliff of 30 airmen who stood to attention for "taps," the final salute to the American Flag and took charge of the flag-lowering event. Then they moved to one side and Leading Seaman Walker and P.O. Roberts of RCN, took charge of the Canadian flag and Sgt. John Browne and Sgt. Pat Dunne, of the Newfoundland Constabulary, on behalf of the Newfoundland Government, raised the Union Jack and the event was concluded.
>
> Standing to attention for the first event was Captain Billy Shannon of Pepperrell AFB, Commander Bremner of RCN and Ray Manning of Department of Public Works. The latter two stood together for the final take-over as the flags were raised and were afterwards joined by Captain Shannon, all three men moving off the flag ramp together to signify that the turn-over of the base to the joint federal-provincial board pending the decision on ownership of the Supreme Court had been completed.
>
> Also present for the USAF were Lieut.-Col. Tom Corrigan of Goose Bay and Col. Webster, Stuart AFB, N.Y. The latter was one of the "originals" who came to Newfoundland in the UST *Edmund B. Alexander* in January, 1941. Chief of Police E. Pittman, and US Vice-Consul J. Blowers [were also in attendance.]

Early in 1960 when it was first learned that the base would close in a year, the American Legion, Fort Pepperrell Post 9, requested the Base Commander to insure they were given the last flag to fly when it was taken down. In 1961, by Act of Congress, the flag was awarded to the American Legion Post. Capt. Billy Shannon, Base Commander at the time of Pepperrell's closing, presented the flag to the American Legion on 15 August 1960. The flag is now on display at the American Legion, in their memorial case.

Immediately after the official closing of Pepperrell and all the buildings and areas, a caretaker force of about fifty, mostly civilians, continued to secure buildings and maintain vital ones. All the 208 buildings and outside area facilities were sold to the Government of Canada in exchange for $1.00. A federal-provincial Board was established and the buildings and areas were disbursed. Most of the facilities west of Virginia Waters were to be designated for federal use, and most east of Virginia Waters were designated to the provincial government. Over the years some of the structures have been sold by the provincial government to private enterprise, however, the majority of the buildings and areas are in use by the provincial government. The federal government still maintains all of its buildings. Today, mostly all the outside of the structures have been painted pastel shades, some new apartment buildings erected, barracks converted to apartment units, and some additions and alterations have been made to some structures. Nevertheless, the base and its fond memories were once again made a reality for over 500 former Pepperrell officers and enlisted men who returned to the "Yank! Come Back to Newfoundland" campaign in 1988, and to the hundreds more who returned in 1989. The campaign was a part of the "Great '88 Soiree," and sponsored by the American Legion, Fort Pepperrell, Post 9.

Group photograph of Headquarters Battery, 421st CA (BN Composite) Signal Hill, summer 1941.

4

Gander Airfield

After the Declaration of War by Britain on Germany in September 1939, the need to strengthen British air striking power became imperative. The manufacture of both bomber and fighter aircraft in England began to escalate. Emphasis, however, was placed on fighter aircraft required for the immediate defence of Britain in the event of a German bomber attack. The first air attack on England occurred on 8 August 1940.

In the meanwhile, Canada was not capable of mass manufacture of bomber type aircraft to meet the demand. The United States was, and immediate arrangements were made for the procurement of US-made bombers. Delivery of US war planes flown across its border was in violation of the US Neutrality Act, therefore the only way to get the planes to Canada, and then to England, was to fly them to US bases in Maine and North Dakota, which were right on the US-Canada borderline. To bring them legally across to Canada, teams of horses or tractors were used to tow them across the border. This system was used from September 1939 until after 14

United States neutrality laws prohibited the flying of military aircraft out of the United States to any foreign country. To circumvent this law, bombers were flown to the US-Canada border, then taken across by horse teams or tractors. Here a bomber is taken across the border at Pembina, North Dakota.

December 1941. Thousands of US-made Lockheed Hudson bombers were delivered to Canada by this method.

At first it was determined that the only way to get the Hudson Bombers from Canada to England was to dissemble the wings and other major parts and transport them across the Atlantic by ship. It was considered unsafe to fly them to their destination because of four main reasons: there was no fully equipped airport facility established between Canada and England where the aircraft could be refuelled and serviced; there were very few air navigators trained for such flights; the dangers existed of an encounter with the German *Luftwaffe* or German Navy; and there was a lack of adequate navigational aids and meteorological services. The method of sea transport for the aircraft continued until late 1940.

In June 1936 construction began in central Newfoundland, in an area known as Hattie's Camp, on an airfield for the Commission of Government. The airfield was to be known as Newfoundland Airport, Gander. On 11 January 1938 the first aircraft, piloted by Captain Douglas Frazer, landed at Gander. By 1939 the airport had one large hangar, an administration building and some small support buildings. Some contracting temporary sheds were erected on site. The one runway was long enough to take care of medium sized aircraft of the time, but no concrete or asphalt was laid until the latter part of 1939.

The problem of expediting the shipment of the US-made Hudson Bombers from Canada to England was considered critical in the spring and summer of 1940. Lord Beaverbrook of England appointed Capt. D.C.T. Bennett of the Royal Air Force to head the newly formed Ferry Command, and requested he study the feasibility of transport of the bombers by air to Britain. Thousands of men from Canada and the United States were being trained in Canada as pilots. From 1939 to 1941, 5,500 young Americans crossed the border into Canada to take pilot training. America was not at war at the time and these Americans were too young to join the United States Army Air Corps.

The Royal Canadian Air Force took over the Gander Airfield activities in early 1940, and immediately began to lengthen the existing runway and construct others. They quickly constructed one hangar and a control tower for operational use, and completed the paving of one of the runways. They surveyed the area and drew up plans for an RCAF station at Gander. There were no roads connecting this remote area with any part of the Island of Newfoundland. The only way in or out, besides aircraft, was by the Newfoundland Railway train.

In late October 1940, Capt. Bennett had selected seven Hudson Bombers, with seven individual crews consisting of twenty-two men, to venture the experimental flight of the bombers across the

Atlantic to Britain. Gander was selected as the base from which the experimental flight would begin. Early in November 1940 the seven bombers and their crews landed at Gander to begin final flight preparations. There were no housing facilities available for the crews, so prearrangements were made with the Newfoundland Railway to place three Pullman sleeper rail cars and one diner car on the rail siding at Gander Railway Station.

On 10 November 1940 the Atlantic Ferry Command began its renowned career when the seven unarmed Hudson Bombers took off, and landed ten hours later in Aldergrove, Ireland. It was the first successful flight for these heavy aircraft, using the Great Circle route to reach their destination. The Atlantic Ferry Command was born. Air Chief Marshall Frederick Bowhill of the Royal Air Force was appointed Commander of the newly formed Ferry Command. By this time the United States was preparing for the production of great numbers of other type bombers, such as the B-24 Liberator and the B-17 Flying Fortress. Canada was manufacturing the well-known Lancaster Bomber as well as some Liberators. A large number of fighter aircraft was coming off the assembly lines, however, both the fighter and short-range bombers did not have the range to fly, non-stop, from Newfoundland to Britain. They had to be shipped by sea.

When the United States entered the war in December 1941, all defence plans changed dramatically. There were no neutrality restrictions now, and the US could send the bombers and fighters directly to Great Britain. The United States was to fight the war on two fronts, the European and Far East. It was agreed by the Allied Supreme Commanders that Germany would be conquered first, providing the United States could hold off the Japanese for a period of time.

Members of Captain Bennett's crews, posing alongside the Newfoundland Railway cars, where they were quartered in 1940, prior to their departure across the Atlantic in their seven Lockheed Hudson Bombers. Total personnel consisted of nine Americans, six British, six Canadians and one Australian.

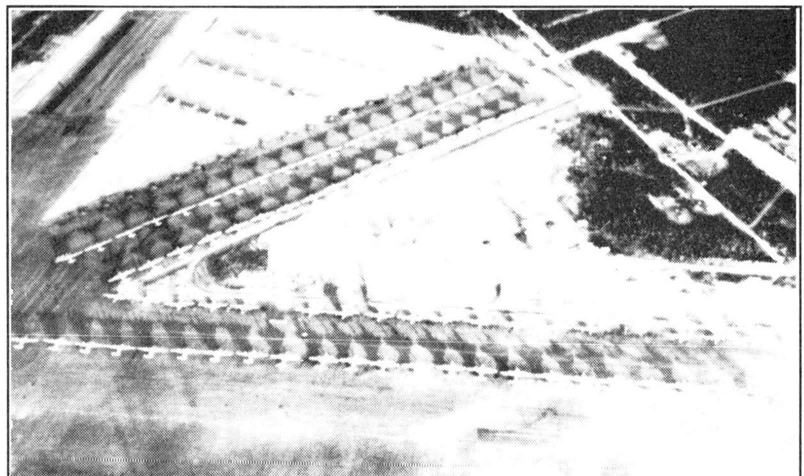

Over one hundred B-17 Flying Fortresses and B-24 Liberator Bombers line up wing tip to wing tip on the runways and parking apron at Gander in 1943. All these aircraft were en route to England via the RAF and USAAF Ferry Command. More military aircraft moved through Gander from 1942 to 1945 than any other airport in the world.

Gander Newfoundland Railway Station, Building 140, was erected in 1939. Despite the fact that it has new siding and other renovations, the building still maintains its original look.

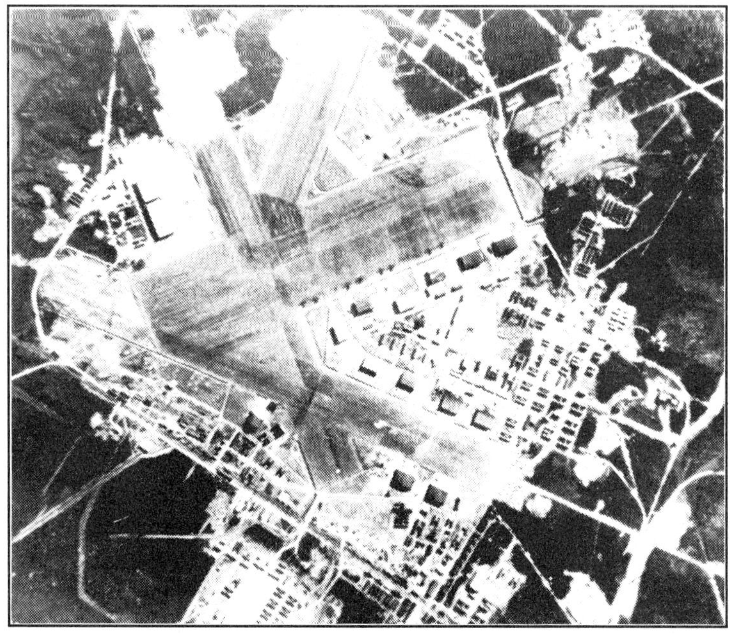

Gander Airport, early 1942. The arrival of US troop ships and aircraft troop carriers by 1944 brought 4,000 assigned personnel to Gander.

Six of the seven Lockheed Bombers that left Gander on 10 November 1940 en route to England.

All the US bases in Newfoundland were placed on full alert. Activity at each was dramatically increased. Harmon Field was not yet ready for trans-atlantic shipment of aircraft. Argentia Naval Station was operational, however, it had its own mission to fulfill. The RCAF stations at Torbay Airport and Goose Bay, Labrador had just started. There were no Allied air bases between Newfoundland and Great Britain at that time. The decision was made by the Permanent Joint Board on Defence to concentrate on making Gander and Goose Bay the two main stopover bases from mainland Canada and the United States, to fly the bombers across the Atlantic. Harmon Field was to be included as soon as its runways were complete, which was not until 1942. The major problem of flying fighter aircraft and short-range bombers to Great Britain was solved when the United States started to build US air bases in Frobisher Bay, Greenland and Iceland in 1942. The stepping stone route was now established. Heavy and medium bombers could fly direct from Gander or Goose Bay. Fighters and short-range bombers could fly from Gander or Goose Bay to Greenland, Iceland and Great Britain. Frobisher Bay (Crystal 11), established primarily as a weather station, was used as an alternate for aircraft developing engine trouble. The route would be via the Great Circle to the north. Full aircraft crews were now assigned to each armed aircraft as they left the US or Canada for Gander. Those pilots who were in the Ferry Command would return to the US or Canada by the Great Circle to the south, via Bermuda. These flights, carrying pilots and crews back for more planes, were made by Catalina Flying Boat type aircraft and modified B-24 Liberators. By the end of 1941, Gander airfield was a beehive of activity.

The Permanent Joint Board on Defence, realizing in 1941 that the escalating war in Europe was placing an extreme demand on Gander and Goose to keep up with the flying activities of each location, agreed to allow the United States to enter into an agreement with the Newfoundland Commission of Government and the Government of Canada, to construct their own facilities at Gander and Goose. Things were rather slow in starting until 14 December 1941 when America entered the European War. The end of 1941 and beginning of 1942 saw a major change at Gander when the Americans moved in and constructed their own "American Side" of Gander, next to the RCAF facilities. Additional hangars and over 100 other types of temporary buildings were erected. The UST *Siboney* arrived in St. John's in October 1941 with over 150 United States Air Corps troops destined for Gander. They were transported to Gander by the Newfoundland Express, set up as a special military train. They were the first US troops to be stationed in Gander.

The RAF Ferry Command consisted of pilots and crews from Britain, United States, Canada and Allied countries. Dorval Airport in Montreal was the first originating base of all bombers to be flown to England, and later to Poland and Russia via Gander. The aircraft and their crews would leave Montreal for Gander and after refuelling, they would continue on their journey. Later on, bases were established in the United States, like Presque Isle, Maine, and others on the eastern seaboard, to maintain the continuous flow of aircraft to Europe via Gander and Goose Bay. Over 10,000 aircraft were flown to England and Europe during this time, by both the American and RAF Ferry Commands.

It was not uncommon to see about 100 bombers on the horizon, which would circle and land, one at a time, at Gander. After a few hours of refuelling and servicing the aircraft and feeding the crews, they would take off and head east. As soon as they were gone, one could see another formation of incoming aircraft on the western horizon. When the bases in Greenland and Iceland were ready, the fighters and short-range bombers started to arrive. They would refuel and take off on their stepping stone route to Goose Bay, Narsarssuak, or Sondrestrom bases in Greenland, then to Kaflavik, Iceland, and off to Preswick, Scotland, or another destined airfield in Britain.

Gander Air Base was the main station used to relay navigational and meteorological information to all the bases in Labrador, Baffin Island and Greenland. All aircraft took their bearings from Gander. The United States Army Air Communication System (AACS) arrived in Gander in 1941. Twenty enlisted men and two officers made up the eight Airways Communications Squadron, who operated out of a truck on the runway until a proper Direction Finder transmitter was set up in Hangar 3. Later on, as new equipment arrived, and much needed additional space was found, they moved into a separate location a few miles from the base. Tens of thousands of aircraft moved in and out of Gander from 1941 to 1944. One can readily imagine the job of the AACS in keeping track of all these flights and furnishing crews with vital weather and navigational data. The AACS personnel also kept track of all aircraft over Newfoundland while they were en route from the United States or Canada to Atlantic destinations. They worked very closely with their RCAF counterparts.

About the end of December 1941, the United States Air Transport Command was organized. Prior to that, it was known as the USAF Ferry Command, which was operating as a separate unit within the RAF Ferry Command Operation. Its mission was primarily the movement of troops and war supplies to Britain and Europe, as well as engaging in the ferrying of aircraft for the newly formed British-based US Eighth Air Force. In one month alone, from one of the ATC bases at La Guardia Field in New York, they moved through bases in Newfoundland over 500 planes, 4,000 passengers and 2,500 pounds of air cargo. In 1944 the Wing Headquarters moved

The first American gas station at Gander, January 1942. Here, soldiers from the 231st Quartermaster Company are refueling a military vehicle: Sgt. Wallace McKeever tipping the drum, Pvt. Arthur Oswald using gas hose, Pvt. Anthony Mazzio looking on. All are from New York State.

The Gander beacon was used to guide incoming aircraft.This structure also contained large air raid sirens.

Mobil Ground Control approach used by the US Army Air Corps at Gander in 1942.

Pilots and crews of the 432nd Reconnaissance Squadron, USAAF stationed at Gander 1942-44.

to Grenier Field, New Hampshire and then to Dow Field at Bangor, Maine. Although Ernest Harmon Air Base was the primary stopover in Newfoundland for the Air Transport Command, many of the flights were diverted through Gander, thus adding to the incredible number of aircraft that had to be serviced before proceeding to their final destination.

The number of Newfoundland civilians employed at Gander during the war years is estimated at 1,500. Housing facilities had to be provided for them because there was no town site in those days, and no nearby communities where the civilians could stay. They enjoyed many base privileges as there were no local stores available. Food services and housing requirements, such as supply, janitorial services, billeting offices and other administrative assignments were performed primarily by the civilians. They were also employed as technicians, engineers, meteorologists, administrators, carpenters, electricians, clerks, secretaries, maintenance, vehicle mechanics, heavy equipment operators, masons and labourers. Civilian construction workers lived in temporary housing erected by the general contractor, and they were located along the railway tracks outside the base area.

Gander Airfield and its operations were wrapped in complete secrecy during the war years. Local newspapers and magazines printed in Gander for the general information of all the military concerned had the words "US Army Base - Somewhere in Newfoundland." Military personnel were requested to keep the word "Gander" out of their correspondence. The installation was known only as APO 865 (Army Post Office). Dependents of US military personnel from the States were not allowed to join their husbands, as they were authorized to do at Fort Pepperrell and Harmon Field in 1941. Those US soldiers who married Newfoundland girls had a hard time getting together, since there

This structure, Building 92, was once the pump house, supplying water from Gander Lake to the airport.

were no married quarters on the base. The usual service clubs were organized, such as the Officers' and NCO Mess, and the Special Service Club, which was frequented by enlisted men of all ranks. Because of the isolation involved, Gander was not considered as an ideal place for duty. However, like the military at other isolated locations, they made do with what they had and enjoyed their off-duty time as best they could. The RCAF operated Radio Station VORG and was able to bring in many programs from the United States by special hook-up.

At the end of the European war in May 1945, the remaining serviceable American and Canadian aircraft involved in the war were scheduled to return. American aircraft, upon arrival in the States, were immediately renovated, up-dated and made ready for the continuation of the Japanese war in the Pacific. Thousands of these planes were returned via Gander and Goose Bay, with Gander processing the most. Many of these aircraft were damaged in the war and patched up for the return flight home. Many did not make it, and Search and Rescue units from Gander and Argentia Naval Base were constantly on the alert. Many emergency landings were made with engines out, or on fire, landing gear problems, flaps or tail sections damaged and hydraulic and electrical systems not functioning properly. Flight line crews, especially crash crews and ambulances, worked around the clock to keep the runways free of disabled aircraft.

Around the middle of the summer of 1945, many United States Army Air Corps personnel were returned to the States for discharge or reassignment to other duties overseas, particularly in the Pacific. The return of aircraft from Britain and Europe continued up to the last of 1946 and Gander's military strength was kept at a maximum until then. By 1947, very few US Army Air Corps personnel remained at Gander. The mission of Gander as a staging point for the RAF and USAF Ferry Command operations was now beginning to be history.

The old administration building at Gander, with some of the military barracks on the left. Hangar 20 is on the right. Also shown is the large area of paved runways and aprons.

These structures were used by the Americans as the Base Exchange and storage buildings. They are still in use today, although they show their forty-eight year age.

In 1948 the American side of Gander was closed. All of the buildings and installed equipment were turned over to the Government of Newfoundland. The RCAF operation also phased out around this time and only skeleton crews remained to look after the occasional military flights passing through. By the end of 1949 many of the military buildings were demolished. The Canadian Department of Transport maintained the airport. The first Trans Canada Airlines (TCA) flight into Gander was on 1 July 1942 and the airport was opened in 1945 to all trans-atlantic flights. It was a designated refuelling stop with passengers and crew resting facilities. By the end of 1950 Gander was known as the Crossroads of the World, the number of commercial overseas flights in and out making it the busiest international airport in North America.

Some of the US military buildings left after 1948 are still in use today at Gander. During the tenure of the Northeast Air Command and the 64th Air Division Commands, many US Air Force personnel were stationed in Gander to look after military flights originating in and out of the area. When the USAF bases at Pepperrell and Harmon closed, a handful of military personnel remained there. Today, there are sixteen or more USAF personnel stationed in Gander, who look after the USAF interests whenever their planes land there.

The United States Service Organization (USO) included Gander in all the visits made by movie stars to Newfoundland and Labrador. The first visit was made in 1943 by Bob Hope, Frank Sinatra and Francis Langford. Several well-known bands, like Les Brown and his Band of Renown, visited Gander and performed for the troops and played for special dances. When the magnificent USO building opened in Corner Brook in 1944, three-day passes were a common

The old USAF central heating plant used to heat the buildings on the base. The facility had a third floor, which has been removed. Renovated, it is now used by Transport Canada.

request by the soldiers anxious to go there. They were authorized a three-day pass once every three months. The large gymnasium, with all the latest equipment, was a favourite spot for many who wanted a good work-out. The gym also featured an indoor pool, bowling alley and large auditorium. The Base Theatre had a new movie every two days and featured top movies of the time. The Officers' and NCO Mess, as well as the stocked Post Exchange (PX) allowed GI's to buy their favourite items, as well as purchase many things they could send home as gifts. Skiing, fishing, boating, hiking and summer sports were all part of the soldiers' recreation. The USAAC had their own band and could cater to squadron parties, weekend dances and special occasion performances. Some soldiers would take a three-day pass in St. John's or in Grand Falls, some sixty miles west of Gander. The US Signal Corps Repeater Station at Grand Falls offered them lodging and meals at no cost. While in St. John's they could stay at the USO or the transient barracks at Fort Pepperrell. In many cases, night life and overall recreation at Gander Air Base were lively. Over 500 Newfoundland girls worked on the base and lived in their own female barracks. Civilian employees were invited to the Officers', NCO's and Enlisted Men's clubs and other facilities.

After the war, Gander Airport underwent another major construction program to convert all facilities to commercial flying. Runways were lengthened and new airport facilities constructed. The name "Newfoundland Airport" was replaced by Gander International Airport. Many families moved into the area and hundreds were employed at the International Airport. In 1954 Gander was given Town status. Gander International Airport still maintains its status, with aircraft from all nations on trans-atlantic flights, stopping over. Even the non-stop flights from Britain, Europe, Canada, United States and others usually fly over Gander for their bearings. Commercial flights to points within the province and mainland Canada operate on a continuous basis.

Above, a 1987 photograph of the Town of Gander. Incorporated in 1954, its current population is 12,000.

Below, Gander International Airport , 1987. This scene shows Air Canada and Russian Aeroflot aircraft in front of the terminal. Note the two radar domes of the AC&W Station on the centre left.

To commemorate the history of Gander and its important role in the European war, a memorial to the Atlantic Ferry Command is erected on Skipper's Hill. It is a Lockheed Hudson Bomber of the same vintage as the one that first flew out of Gander on the first trans-atlantic crossing in November 1940. Over 10,000 Hudson Bombers were ferried through Gander from 1941 to 1944.

Gander has the only War Graves Cemetery in the province. The following is an excerpt from an article by Amy Louise Peyton, "The Unsung Heroes of the Ferry Command" as it appeared in the *Senior Citizens News*, 1988:

> During the early war years, the new airport and surrounding area were the scene of many aircraft crashes. A burial ground for the victims was created to one side of the airport runways. As the casualties mounted, a "war graves cemetery" was prepared. At this spot, far from grieving relatives and friends, lie 101 "unsung heroes." They had a dream, a dream to help restore peace; to them an earthly dream unfulfilled.
>
> The Commonwealth War Graves Commission cemetery at Gander is unique, the only one of its kind in the Province. Carved out of the wilderness, atop a level plateau overlooking Gander Lake, the lawned graveyard is unfenced but well defined. In its natural setting it is surrounded on either side by a dense, almost impenetrable forest of spruce and fir. The neat rows of headstones in this well-cared for cemetery are perhaps not as serried as those of the battlefields but the young ages are the same. A cross of sacrifice stands as a lonely sentinel, arms outstretched, amidst the "flowers of youth."

Gander War Graves Cemetery. Many American servicemen who died in various accidents in and around Gander from 1941 to 1945 were buried there. In 1946-47 they were exhumed and returned to the United States for burial with full military honours.

One of the Hudson Bombers used in World War Two, and presumably flown to England via Gander during the war. The aircraft was a gift to the Town of Gander and is mounted in a most prominent place for public viewing.

A get-together of the US Army Air Force Weather Squadron and office staff, 1942. This Weather Squadron was assigned to the 429th Bomb Squadron station at Gander Airfield on the American side.

5

Harmon Field Stephenville

The largest area specified in the Anglo-American Lend-Lease Agreement was designated at Stephenville, Newfoundland, for the construction of an Army Air Force Base. The total of 8,159 acres, selected in October 1940, was situated at the northeast end of St. George's Bay. This was to become the largest military airport of the United States Army Air Force outside the continental USA. It was to be known in 1941 as Stephenville Air Base.

Stephenville was a small, quiet village up to the beginning of 1941. The 500 residents were engaged in farming, lumbering and fishing. Their homes were scattered all over the area onto which the huge military installation was to be constructed. The United States was not actively involved in the war, therefore, in the beginning, some of the residents could not understand why they would lose their land and homes, but arrangements were made for compensation. In 1941-42, the community of Stephenville was transformed into a boom-town when, almost overnight, the population increased to over 7,000 people, most of whom were engaged in the base construction.

Because of its strategic location, and the fact that flying conditions were excellent over 90 percent of the time, Stephenville Air Base was considered to be a vital refuelling stop for aircraft en route to Europe. After the United States declared war on Japan on 7 December 1941, the Military Chief of Staff immediately knew that Stephenville Air Base alone could not look after the thousands of aircraft that would be flown to the European Theatre of War. The US Army Air Corps Base became a part of the Newfoundland Base Command from its beginnings.

The first contingent of the US Army troops arrived in Stephenville in January 1941, consisting of US Corps of Engineers

Harmon's "main drag," taken in early winter 1942. Tent city had closed by this time at Camp Morris and all soldiers and most civilians who worked on the Base were moved into their temporary quarters.

Camp Morris at Stephenville, Harmon Field, 1942.

One of the new civilian barracks constructed in 1948. These dormatories were the same as those constructed for the Military.

The old Air Force Post Exchange (PX). This was replaced in 1948 by a more modern store to serve the men and women assigned to tbe Base. Due to Customs regulations, Newfoundland civilian personnel were not allowed to purchase articles in the PX .

and key US civilian personnel. There was only a handful of them, however in a short time they had completed their surveys of the area and determined what would be constructed. In February 1941, about 150 members of the US Signal Corps and a small contingent of the 24th Coast Artillery came ashore from a transport anchored in the harbour. They set up tents and bivouacked near the town. Temporary buildings were erected in strategic locations and 155-millimetre guns were placed around the perimeter of Stephenville Air Base. A company for the First Battalion of the United States 3rd Infantry, part of the original contingent aboard the UST *Edmund B. Alexander* was also assigned there to assist other US Army personnel. They performed security duties in and around the whole area under US military contract. All of these troops were transported from St. John's to Stephenville Crossing by the "Newfoundland Express" train. There was no railroad branch between Stephenville and Stephenville Crossing (about six miles away). The US Infantry personnel completed their journey to Stephenville by road. The first Base Commander was Major John A. Gaven. The mission of the US Signal Corps was to set up a communications network from Fort Pepperrell, to Stephenville, to the United States.

The Hotel DeGinque was used as a dormitory for transient officers and visiting dignitaries. It is now Hotel Stephenville.

The Newfoundland Base Contractors, working under the US Corps of Engineers, awarded its first contract to an American contracting company in February 1941, for the construction of an airfield and all of the component buildings necessary. Several of the former residents' houses, and buildings that could be relocated, were used by the Corps of Engineers and Contractors as offices and living quarters. The original contract called for all temporary structures. Even the hangars fell into this category. However, this was changed after December 1941 to semi-permanent hangars. The permanent structures came some time later. Over 1,500 Newfoundlanders worked on construction of the base.

On 23 June 1941 Stephenville Air Base was officially named, by Act of Congress, as Harmon Field, in honour of Capt. Ernest Emery Harmon, a pioneer in United States military aviation history. Capt. Harmon was born in Dallas, Texas on 8 February 1893. After receiving flight training, he was commissioned a Second Lieutenant on 5 April 1918 and received an appointment as Second Lieutenant in the regular army on 1 July 1920. Acting as test pilot for the United States patent office, Capt. Harmon was selected to pilot the Martin Bomber in the "round the rim" flight of 1919, which circled the boundaries of the US for the first time. While making a test flight from Maryland to Mitchel Field, New York, on 27 August 1933, Captain Harmon lost his life when his aircraft ran out of fuel near Stamford, Connecticut.

The base was not ready on 7 December 1941 when America declared war on the Japanese. The Newfoundland Base Contractors were told to leave all buildings occupied by them in place. These were scheduled to be demolished when the base was completed. The US Army Air Force (USAAF) personnel were quartered on the base in tents and other temporary structures until new quarters were completed. Harmon Field was not completed when in April of 1942 over 700 US Army Air Corps personnel, including the 429th Bombardment Squadron, arrived by military sea transport. It was necessary for these troops to stay aboard the military transport until their tent accommodations were ready. The largest "tent city" to that time to be erected in Newfoundland, consisting of over 100 mammoth size tents, to temporarily accommodate these troops, was called Camp Morris. There were very few organizations at Harmon at the time, including the Army Coast Artillery, Army Transport Command, Signal Corps, and the Army Air Communications Squadron. However, there was a Port Company, Trucking Company, Ordinance Company, and Quartermaster along with medics, bakers, cooks and others to make up a troop contingency. By April 1943, there were seventeen military units assigned and more than 4,000 US soldiers.

Although it appeared in late 1942 that Harmon Field had all the facilities and personnel in place for operations, it wasn't until August 1943 that it was opened for heavy air traffic. Prior to that time and from early 1942, the runways were used only for emergency landings. From early January 1942, B-17 bombers used the incomplete runway, where temporary field erected radio, tower and radio range facilities were used. The first Army Air Communications Squadron detachment of five men was set up at Harmon on 15 February 1942, and the first communication with Gander Air Base was transmitted on 3 March 1942. A detachment from the 8th Weather Squadron also set up at Harmon about the same time. The first non-emergency aircraft to land at Harmon was an RW07 on 15 September 1942.

In September 1943 Harmon Field, although remaining under the Newfoundland Base Command, transferred to the Army Air Transport Service. Its mission was to service all aircraft involved in the air movement of personnel and supplies to the European Theatre. In addition to hundreds of C-47 (DC-3) and C-54 (DC-4) type aircraft in 1943-44, it also catered to the needs of B-17 bombers, B-24 Liberator bombers and various fighter-type aircraft and many more. Three B-17 bombers were assigned to Harmon Field from 1942-44 and used extensively in the patrol of the Atlantic for German U-boats.

The end of the war in Europe increased the activity at Harmon Field with the movement of US troops returning to the United States. The "Skymaster" was then in use and it was not uncommon to see fifteen or twenty of those huge aircraft on the apron. Crews and passengers were fed and rested at Harmon for the final lap of their long flights. While the US Air Bases at Goose, Gander and Argentia looked after the return of both United States and Canadian aircraft from the war zone and England, Harmon Field concentrated mostly on the movement of personnel and supplies.

A major construction program began at Harmon in 1947 to construct permanent buildings in place of the temporary ones built before and during World War Two. The purpose was to improve and expand the base so that it could eventually become a permanent overseas United States Air Force Base. The construction program continued until 1950 and hundreds of civilians were employed in this new conversion program. Runways were extended to take care of the new aircraft being developed, and many more up-to-date service buildings were erected. Military and civilian personnel moved out of their old quarters and into new ones as soon as they were completed. By the end of 1950 the new United States Air Force Base was one of the most elegant of any overseas military installations.

During the latter part of 1945, most of the US Army troops were deployed to the United States or to new assignments in the Pacific. On 1 July 1948 Harmon Field was transferred from the jurisdiction of the United States Army Air Corps (USAAC) to the newly formed United States Air Force. All personnel and equipment came under the Military Air Transport Service (MATS) of the Air Transport Command. Each year, up to 1965, over 4,500 trans-atlantic flights stopped over at Harmon. For that seventeen-year period, each year 132,000 passengers and 3,000 tons of air cargo passed through Harmon Field.

On 1 July 1948 Harmon Field was renamed Ernest Harmon Air Force Base. Approximately 2,000 military personnel were stationed there during that year. In 1953, like all other bases and sites in Newfoundland and Labrador, Ernest Harmon became a part of the Northeast Air Command. Its mission was to participate in the supply and servicing of all US Installations in NEAC, including American bases in Greenland and Baffin Island. Ernest Harmon retained its importance as the first major overseas stop for military aircraft flying

A view of Main Street, Harmon Air Force Base, in 1948. All of these temporary structures have since been removed and new, permanent buildings have been erected on the site. This road leads directly into Stephenville through the main military security gate.

A KC-97 Stratotanker–38 feet high, 117 feet long and 141 feet from wing tip to wing tip. It carried a crew of 5 and 41,000 gallons of fuel. They were stationed at Stephenville.

the North Atlantic to Greenland, England and Europe. The Strategic Air Command (SAC) operated out of Harmon from 1953 to 1958.

In 1953 Harmon AFB once again underwent another major construction program. Runways were lengthened and several new buildings were built, including a $2 million hospital with a 100-bed capacity. The well-known "Black Hangar" was moved from its original location to the lower east side of the new aircraft parking area. Aircraft parking aprons were widened and the so-called "Boon Docks" were constructed to take care of the maintenance and servicing requirements of large aircraft. Underground refuelling lines and hydrants were installed on the apron. Harmon became the home of the renowned 323rd F-102 Flight Interceptor (Jets) Squadron. Over forty KC-97 "Stratotankers" of the 376th Air Refuelling Squadron of the 4081st Strategic Wing under the Eighth Air Force, SAC, were also assigned to support the jet fighter squadron. All the activity was part of the air defence against a possible cross-polar attack from Russia. It was all part of the American-Canadian NORAD defence network. This activity grew rapidly under the 64th Air Division, when it took over from the Northeast Air Command in 1958, and continued to grow under the Eighth Air Force, Strategic Air Command, until the base closed in 1966. Command Headquarters for the 64th Air Division was located at Pepperrell Air Force Base, however, Ernest Harmon had its own defence mission. Over $10 million was spent on construction from 1958 to 1966. It was a most common sight around Stephenville skies to see two or three formations of F-102 jets or ten or more giant KC-97 supertanker aircraft manoeuvring, before jointly departing to some rendezvous over the Atlantic to meet other American aircraft and refuel them in the air.

The arrival of supplies by military sea transport for construction of the base presented a problem. There were no deep-water docking facilities, and much had to be brought ashore by boats and barges. The US Army built a railway branch from Stephenville Crossing to Stephenville in 1942. After that, a lot of material was shipped from Port aux Basques for transport by rail to Harmon. The engineers built a road to Stephenville Crossing. The 156th

Above, an interior view of the old control tower at Harmon around 1947. Note the old-style equipment as compared to the computerized equipment of today.

Below, part of Flight Operations Flight Centre in 1941. All incoming and outgoing aircraft were noted on these boards. There were no computers in those days.

Above, a 1960 aerial view of Harmon showing part of the central core of the Base. Housing units can be seen in the top left. The two seven-storey Harmon Hilton dormitories can be seen in the centre right.

Below, a 1960 close-up aerial shot of the elaborate Officers' and NCO married quarters at Harmon.

Transportation Post Company, who arrived in 1942, first constructed the harbour at Harmon to facilitate the unloading of troops and military supplies needed for construction.

The second United Service Organization (USO) building to be erected in Newfoundland was located between West and Park Streets in Corner Brook. In July 1943 a contract was let to Bowater's Newfoundland Pulp and Paper Mills Limited to construct the facility. After months of delay in trying to obtain building materials, it was finally completed and opened on 21 February 1944. The opening was a gala affair with local government and Corner Brook residents, and the United States Army Air Corps Band in attendance. The US Government Administration Office recreation building was to serve the soldiers from Harmon Field, Gander, Stephenville Crossing, Howley, Millertown Junction and Grand Falls where the US Military were stationed. It had a most attractive lounge with a large fireplace of red tapestry brick. Off the lounge was a quiet library, writing room and reading room, mock bar, club room, a fifty-two man dormitory, a large kitchen and dining room. The USO building served US Military personnel on pass to Corner Brook until it was sold in 1946, and afterwards was known as the "White House."

The USO arranged for many movie stars to perform at Harmon and at the USO building in Corner Brook. Some of them were Joan Blondell, Bob Hope, Frank Sinatra, Jayne Mansfield, Lana Turner, Elvis Presley, Frank Lovejoy, Raymond Burr, Forest Tucker, Marlene Dietrich, Phil Silvers, Ed Sullivan and Edward G. Robinson. Over the years, Harmon played host to many visiting dignitaries such as Her Majesty Queen Elizabeth II, Prince Philip, former Premier Joseph Smallwood, Franklin D. Roosevelt, John Diefenbaker, Dean

The "White House," the USO building in Corner Brook, opened in 1944. The structure was sold to private enterprise in 1946. It is still in use today.

Acheson, Prime Minister Nehru of India, Conrad Adenauer, and Dag Hammershjold. Some of the visiting top military leaders were General Dwight D. Eisenhower, General George Patton, Colonel James Doolittle, General Omar Bradley and General Hoyt Vanderberg.

The GI Trolley will long be remembered by Harmon soldiers and their friends from Stephenville, as it was a symbol of good times off the base. It represented practically the only transportation link with the rest of Newfoundland. Built in 1943 by Harmon soldiers, the Trolley was a model LS truck equipped with a Hercules 6-cylinder gas engine and gauged for the narrow, 32-inch rails of the Newfoundland Railway. The Trolley, or PM-4, could accommodate forty-five passengers and made the trip to Corner Brook three times a week. The original operators, Sgt. Rudolph Maynard and Cpl. Hubert North, were the engineers on the trolley and Mr. J.J. Penney, the Newfoundland pilot. The GI Trolley operated until the fall of 1945; the pride and joy of the American GI.

Over the years, Harmon Air Force Base provided a variety of recreational areas and clubs for the relaxation and enjoyment of the US Military and Newfoundland civilians. During the summer months, Special Services provided regular three-day trips to Bottom Brook Camp. Those involved would travel to Stephenville Crossing by truck and then by boat to the camp. Later on, an improvised recreational facility was set up at Camp 33, some thirty miles from Stephenville. Moose hunting and fishing were the main sports at Camp 33. The Newfoundland civilians, who lived and worked on the base, had their own lounge called the "Caribou Club." Liquor sold in this establishment was purchased through the Newfoundland Controllers, as cheap liquor and beer available on the base was only for the use of the military.

A 1948 photo of the Caribou Club at Harmon, which at one time had a membership of over 1,000 civilians who worked on the base.

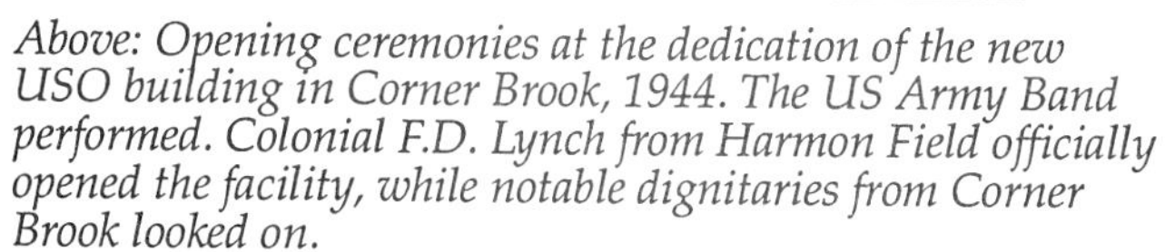

Above: Opening ceremonies at the dedication of the new USO building in Corner Brook, 1944. The US Army Band performed. Colonial F.D. Lynch from Harmon Field officially opened the facility, while notable dignitaries from Corner Brook looked on.

Above right: The NCO Club at Harmon was one of the most modern at any military base outside the US (1953 photo).

Right: Trolley takes on load of US soldiers at Stephenville en route to Corner Brook on three-day passes.

Bottom Right: Newfoundland girls and American soldiers heading for a dance at the USO in Corner Brook, 1944. The civilian on the right is Joe Murphy, a popular USO director.

Below: American Army soldiers on Main Street, Stephenville, with St. Stephen's Church in the background, 1943 .

Prior to 1947 the only recreational services were at the Special Service Club. Mostly military personnel frequented the facility. In 1948 an elaborate Enlisted (Airman's) Club, and in 1950 a luxurious Officers' Club were constructed. The Officers' Club today is occupied by the Royal Canadian Legion; the Airman's Club was sold to private enterprise. A new recreational building was erected in 1953 and has modern facilities. This building and the large indoor swimming pool are now operated by the Town of Stephenville. Large sports fields were developed and competition between Harmon, the other bases and local teams was very common during the summer months. Youth sports and other programs, sponsored by personnel on the base, were continuously conducted from the start of 1941.

The cooperation of the Army, and later the Air Force, to assist in local projects, or when called upon during emergencies, was more than welcomed by the off-base residents. The US Army built the road to Stephenville Crossing and maintained it for a number of years. Stephenville was not incorporated as a community until 1947, and was given Town status in 1951. Up to 1953, none of the streets in Stephenville were paved. The US Military assisted in keeping the main roads open in winter and loaned miscellaneous heavy equipment to the Town on many occasions. When the base fire department was first formed in 1941 and up to 1955, the fire department answered all fire calls in Stephenville, Stephenville Crossing, and as far as Port-au-Port and St. Georges to the west. The United States Army Engineers built the railway from Stephenville to Stephenville Crossing in 1942, and ran its own supply and passenger train from the base to Stephenville Crossing and from Corner Brook. In 1946, permission was granted for dependents of the US military personnel to join them at various Newfoundland bases and hundreds of military families rented houses and apartments in Stephenville and at the "Crossing." A record in 1955 shows there were approximately 1,000 rental units by US servicemen in Stephenville, with a monthly total rental of approximately $165,000 that went into the Stephenville economy. There was almost a complete integration between the local populace and the American servicemen in the Stephenville area, over what a notable historian in Stephenville, Mr. William F. Slade, called "The Carefree Years."

The interior of the bar and lounge are indicative of the elaborate furnishings of the NCO Club. A very spacious lounge for dancing and a games room for slot machines were only part of what the club offered the NCO and his guests.

A great feeling of anxiety hung over the approximately 1,200 civilians who worked on the base, when in 1958 the Northeast Air Command announced it was being deactivated. However, it was learned that the base operations would continue under the new Command of the 64th Air Division. This change in command for Harmon revitalized the base, as a new multi-million dollar expansion program was announced. The two Harmon Hilton dormitories, at a cost of approximately $2 million each, including the centre core, a new hospital and additional Officers' Quarters were only a part of the new construction program. The Strategic Air

Above: The new location of "Black Hangar" and part of the "Boon Docks," 1954.

Below, part of Port Harmon. In the right centre the Newfoundland Linerboard Mill, constructed in 1970, can be seen. A part of the breakwater can be seen in the lower right.

Below left, Base Headquarters used by each major command throughout the active years at Harmon Air Force Base. This was also the main office building of the Harmon Corporation, established in 1966.

Command looked upon Harmon as the nearest refuelling stop. A new range of large military bombers, jet fighter aircraft and in-flight tanker refuelling aircraft came into being. For the next eight years Ernest Harmon Air Force Base flourished as a strategic part of the NORAD Air Defence program. However, all good things eventually come to an end, as in 1965 when the first deactivation of Harmon AFB was announced. Up to the time and from 1957 over 5,000 United States Air Force personnel were stationed at Harmon. The annual operating costs of the base were about $17 million in 1962. Salaries for both military and civilian personnel amounted to more than $1 million a month. The total value of Ernest Harmon Air Force Base, at the time of closing, was estimated at $179 million.

On 16 December 1966, Harmon Air Force Base officially closed and all buildings and facilities were turned over to the federal government. Like at Pepperrell Air Force Base, a Federal-Provincial Board was set up to determine how to utilize best all the buildings and areas. Ernest Harmon was the last base on the Island of Newfoundland to close.

On 26 December 1966, control of all former USAF facilities came under the newly formed Harmon Corporation. Today, the entire complex is part of the Town of Stephenville and the Newfoundland and Labrador Housing Corporation. The airfield is used for commercial aircraft, and is known as Stephenville Airport. It is operated by the federal government. They operate one hangar and some buildings at the airport to take care of airport maintenance and emergency duties. The first Trans Canada flight operations began on 1 September 1949, and both Trans Canada and the USAF shared the then called "Passenger Terminal."

Left, Harmon Theatre, with a seating capacity of over six hundred and a large stage area for various performances, 1960. Many Hollywood stars performed in this theatre. Part of one of the Harmon Hilton dormitories can be seen in the background. Right, the Arts and Culture Centre at Harmon, used by the town of Stephenville. This building was the former administrative headquarters for the Straegic Air Command.

6

Fort McAndrew

In May 1941 construction began on Fort McAndrew which is located immediately adjacent to, and on the south side of, the Argentia Naval Station. Up to the time of occupation by the US Army, the area was known as Marquise. Part of the original 3,392 acres of land allotted to Argentia, as noted in the Anglo-American Lend-Lease Agreement, was to be used for this US Army Base. During the same month, two officers and twenty-eight enlisted men from the 3rd Infantry, 52nd Coast Artillery and 24th Anti-Aircraft Artillery units were assigned duty at the Argentia area. Their mission was to establish temporary gun positions and to provide a 24-hour watch in temporary observation tower posts. They also provided police duties at Argentia, along with the 120 US Marines who first landed at Argentia on 25 January 1941. The main reason for the construction of Fort McAndrew was to provide protection for the Naval Operating Base and the Naval Air Station at Argentia. During the latter part of 1941 a Naval Seacoast Artillery was established and was responsible for harbour patrol, anti-motor torpedo patrols and searchlight positions at Fox Island, Dunville. The US Army Anti-Aircraft Artillery conducted aircraft searchlight protection around the perimeter of the base. The US Infantry conducted motorized and foot patrols with war dogs, and provided a mobile fighting force. In January 1942 the first large contingent of US Army troops arrived by ship. These troops were under the jurisdiction of the Newfoundland Base Command, with their headquarters at Fort Pepperrell. Over 3,000 Army personnel were stationed there during the period of 1942 to 1943. Fort McAndrew was considered, at the time, to be the second largest US Army post outside the United States. (Fort Pepperrell was the largest.)

Besides the mission of security and manning of the large guns protecting the Argentia area, the US Army was engaged in many other activities. It operated a major part of the port passenger and

The coastline of the original Marquise. This photo shows some of the original houses belonging to the former Newfoundland residents in the forefront. These buildings were eventually relocated to make room for the Army's railroad, nicknamed the "Ruptured Duck." The railway can be seen from the pier around the shoreline, and on the Navy side, Argentia. The low building to the right is the Army Infirmary; at the top is Fort McAndrew family housing units. The wharf was mainly a supply wharf, and just to the right are store houses for further dissemination through the military complex.

Salmonier Cove, Fort McAndrew, Little Placentia Harbour, 2 October 1942.

cargo terminal at the Naval Station; and operated and maintained the on-base "Ruptured Duck" railroad. The Army engineers were engaged in the construction of building facilities for the Newfoundland Long Lines system, weather stations, observation towers and stations. In May 1942 they started to rebuild the road from Argentia to Holyrood, and completed the project in December 1942. The US Army maintained this road up to the time of its evacuation from McAndrew. It also patrolled the roads as far north as Arnold's Cove on the isthmus of the Avalon Peninsula, where it had an observation post, and upgraded the road and railway line between Argentia and Whitbourne.

One of the US Army's depots between Colinet and Holyrood was at Camp 4 on the Salmonier Line. The main building, known as the Staff House, for use by ten to fifteen US soldiers, is all that is left of the complex. The red and white building is currently the summer home of Charles Hoddinott, a veteran civilian employee of the US military in Newfoundland. In 1942, there existed a diesel plant, lighting plant, a forge and machine shop, a pump house and a cook house, all on this property.

In 1946 Fort McAndrew came under the command of the US Army Air Corps with its headquarters remaining at Fort Pepperrell. In 1947 all activities transferred to the United States Air Force and it was then renamed McAndrew Air Force Base. Both the Navy and US Army Air Corps, and later United States Air Force, jointly utilized the runway facilities at Argentia. In 1953 McAndrew was incorporated into the Northeast Air Command network and played an important role as the major supply centre for northern bases and sites.

One of the large guns installed in 1941 and mounted on a 360 degree turntable, one of many that protected the Argentia Base. It is still in position today.

The lower centre of this photo shows Battery "F" and the fortifications on Latine Point. Battery "F" had 90mm ATMB guns to protect the Station at Roche Point. In the centre is the Richard Peck *as she lies tied up to the wharf.*

The chapel at McAndrew Air Force Base was the most modern of Air Force chapels within the Northeast Air Command. Many marriages and christenings were performed here.

McAndrew Air Force Base School. All schools at each major base had up-to-date classrooms and equipment.

The former US Army staff house at Camp 4, Salmonier Line.

In 1954 deactivation of McAndrew AFB began and in 1955 the US Navy took over all the facilities. Civilian personnel were transferred to work at other bases within the Command.

7

US Army Repeater and Direction Finding Stations

In 1940-41 there was limited telephone service throughout Newfoundland, and none in Labrador. The service for Corner Brook and the surrounding area was handled through an exchange installed at Bowater's Pulp and Paper Mills Ltd. This service did not extend beyond Deer Lake to the east. The Anglo-Newfoundland Development Paper Company in Grand Falls also installed a telephone exchange and serviced the town and areas west to Buchans and east to Botwood. On both of these small exchanges, private telephones were at a premium and many shared up to ten party lines. There was no service at Gander in 1940 except that established and maintained by the Royal Canadian Air Force. Where it was necessary to have telephone service in growing community centres, the government's Newfoundland Post and Telegraph allowed them to be installed through their very small exchanges, located in post offices or in Newfoundland Railway Station buildings. The lines for these phones had to follow the route to the lines used by Newfoundland Post and Telegraph, through which telegrams could be transmitted. Usually these lines followed the route of the Newfoundland Railway. On the eastern section of Newfoundland, the Avalon Telephone Company held the franchise on the telephone system. Its area was on the main railway line as far west as Whitbourne, Trinity Bay. Places like Shoal Harbour, Clarenville, Bonavista and Trinity had access to telephone only through the Newfoundland Post and Telegraph.

The Avalon Telephone Company in St. John's was the first telephone company in Newfoundland. In order to establish contact with Canada, a radio telephone service was established in 1937, servicing St. John's and the Burin Peninsula. In 1947 Avalon Telephone began an extension to its services and included the Corner Brook-Stephenville area in 1948, Grand Falls area in 1952, and extended as far as Port aux Basques. In 1949, Canadian National Communications established a telephone service at Gander known as CN Tel and later Terra Nova Tel. They installed and operated a telephone service for central Newfoundland (except Grand Falls area), Bonavista, and White Bay, Green Bay and Hermitage Bay areas. Today, the Newfoundland Telephone Company, established in 1970, covers all of the Island with its service.

When the American military first arrived in October 1940, communicatios presented a most difficult problem. There was no way that the military could tap into the system of the day and provide direct communication with its major army base in St. John's and the Naval base at Argentia. There was one alternative; to install its own communication network. The war warranted complete secrecy on movement of troops and supplies within the Newfoundland area. The function of the proposed communication system, known as the Newfoundland Long Line System, was to provide intra-island land line communication between all USAF and US Naval facilities throughout the Island and provide a link in communication with the United States and Canada.

The US Signal Corps, originally 150 officers and enlisted men, arrived by sea transport at Stephenville in February 1941. Their mission was to establish a radio message centre and telephone communication network across Newfoundland to St. John's. The work had to start immediately, regardless of the winter, as this military communication link was crucial. The plan included setting up remote receiver and transmitting stations approximately fifty to sixty miles apart, and following the route of the Newfoundland Railway in doing so. Additional personnel and equipment were requisitioned from the Corps of Engineers at Army Headquarters at Fort Pepperrell, and more Signal Corps personnel came, as warranted, from the United States. One of the first of the ten areas selected was Fort Pepperrell. It was designated as the operating headquarters for the movement of supplies, personnel and equipment to other stations being constructed, by using the Newfoundland Railway. Stephenville Crossing was equipped with a rail loading siding for the flat and box cars of the Newfoundland Railway. Supplies and equipment that came into Stephenville by military sea transport could be easily transported by truck, and later by rail, to the Crossing. The code name on each of these Army Long Line Installations was "Repeater Station." They were constructed in order from Stephenville Crossing, going east, at Howley, Millertown Junction, Grand Falls, Gander, Shoal Harbour and Whitbourne, and ended at the Base Telephone and Communications Building at Fort Pepperrell, which served as the facility control station for the entire system. In 1943 a special transmitter, operating at a special secret frequency, was installed at Snelgrove, near Windsor Lake. This was considered the most Eastern Terminal. To the west, one at Table

Left, Signal Corps crews had to dig out many times on a siding in order to get the rail speeder and trailer back on the main line.
Right, one of the 21st Signal Corps crews, with their speeder, waiting on a siding while a Newfoundland Railway train rumbles past. Speeder patroling may appear to be fun, but it was rough and dangerous. In winter, on these open cars, the cold weather was penetrating.

The US Army Repeater Station at Grand Falls, 1942. The size of each station depended on its function and mission. Most stations had a section of communication lines and poles to maintain from one station to another. They were all located close to the Newfoundland Railway tracks, as each had a rail speeder assigned.

Mountain, outside St. Andrew's, was a sophisticated terminal with direct communication with the United States and Canada. When Harmon Field became operational, the temporary Repeater Station at Stephenville Crossing was moved into Harmon Field. The building at Stephenville Crossing then became a storage depot for materials and supplies for west coast Long Line maintenance.

In addition to the seven Repeater Stations which were manned installations, there were fifty-seven unattended Repeater Stations between St. John's and St. Andrew's, or approximately one station every seven miles. The complete Long Line was constructed by the 21st Signal Corps Company. Each of the Repeater Stations was constructed with concrete block. As the operations and maintenance requirements on the Signal Corps increased, wooden additions were used for accommodation of the personnel and placement of equipment. Each unit was self-sustaining with its own cook, dining room and small recreational room and sleeping quarters to accommodate between ten and fifteen men. There were two maintenance crews, each working twelve-hour shifts, to keep the 24-hour communications network open. The entire Long Line cable system was maintained under gas (nitrogen) pressure. Each station was assigned a gasoline operated rail speeder and a number of rail trailers and, where roads were available, one carry-all and a jeep. A senior NCO was always in charge of each Repeater Station, however, Signal Corps Officers were regular visitors along the route. Since they used the Newfoundland Railway narrow gauge tracks to maintain their part of the communication line, the NCO in charge of the rail speeder was charged with obtaining rail clearance from one point to another, and he had to be very familiar with all railroad sidings and schedules. Lifting the speeder on and off the tracks was simple, however, the trailers could be carrying over a ton of wire or other materials.

A Signal Corps linesman sawing off the top of a pole in preparation to laying the cable. Note the glass insulators used on the cross arms. The terrain features here tell their own story of the work that was involved.

The main radio transmitter (unmanned) to the United States and Canada was installed on Table Mountain at St. Andrew's. In the east, a similar transmitter was installed at Snelgrove, and manned 24 hours a day. The installation of thousands of telephone poles and 527 miles of cable from Stephenville to St. John's, and west to St. Andrew's, was a difficult task for the 21st Signal Corps. A section of cable was also laid between Whitbourne and Argentia Naval Station. The first cable strung was a 100 pair line. The second line of the same capacity was laid in 1943-44. In some instances, Newfoundland Railway freight trains would move the very heavy coils of wire to the nearest siding, from where the men would work. The job then, without machinery in the wilderness, was to move the cables of wire to the speeder trailer. Many times installation crews had to bivouac in the woods, because the snow on the tracks was too much to allow them to return to their camp. This condition was encountered many times at Kitty's Brook on the Gaff Topsails, which is noted for its snow accumulation and storms.

Installing the thousands of telephone poles required was no easy task; all had to be done by hand. The terrain in Newfoundland is very rocky and the going was tough while trying to manoeuvre around it. Crews could be out for weeks at a time, living in tents in the woods, with provisions brought in daily by the Newfoundland Express, a freight train, or the Signal Corps crew on their speeder, This work continued all through 1941 and 1942. Hundreds of military personnel were engaged in the installations. Some civilians were hired, and when heavy equipment was available and could be used, it was a welcome sight. However, the availability of heavy moving equipment was few and far between; in most cases it was manpower and dog team. The winters of 1941 and 1942 were not the best years, and snow piled up to heights of ten to twelve feet in areas where these men had to work. Where the telephone pole line could not follow the Newfoundland Railway line, it was necessary to detour around large hills, boggy ground or small ponds and streams. In the wintertime this meant moving poles and cable up steep slopes and down treacherous ravines to reach the selected spots.

The most difficult installation of all was Table Mountain near St. Andrew's. Here the terrain was such that the remote station had to be constructed on a high plateau. Great difficulty was experienced in trying to get the poles, cable, and building supplies up very steep mountainous slopes. It took the 21st Signal Corps Company three weeks to complete the installation on Table Mountain. Before it was done, one of the crew was injured with a broken leg, and one member froze to death in a blinding snowstorm. Table Mountain was only 2,000 feet high, but to reach the summit the men had to follow a trail that wound six miles around the mountain's side, up dangerous slopes and over fifteen to twenty foot snowdrifts. The trail threaded through passes that sometimes were almost completely free of snow. Light thaws one day followed by sub-zero weather at night covered the snow with glassy ice. At one point on the mountain, the men were only 300 yards from the top, measured straight up, but they still had almost two miles to travel. Each man carried fifty pounds on his back, making two trips a day up the mountain.

When the light supplies were hauled up, the men tackled the heavier stuff. Hitching themselves to tow ropes, they hauled an improvised sled up the mountainside. But when the time came to haul up some 800 pound crates that could not be broken down, the men were stymied. The more they heaved and strained at the ropes, the deeper they sank into the soft snow. The only way the heavy crates could be taken to the top was to use dogs or horses. Because of the deep snow, horses were ruled out. They requested assistance from the US Army at Stephenville and fourteen Siberian Huskies

Top left: Signal Corps speeder/train accident, 1942 when, during a snowstorm, the Signal Corps crew did not hear or see the train coming. One member suffered a broken leg in the accident.

Top right: Soldiers enjoying dinner aboard the diner car. Three of the cars were used for sleeping at St. Andrew's siding.

Centre: Moving materials by hand-drawn snowsled up the steep mountain side at St.Andrew's. Some snow drifts were twenty feet high.

Bottom: The bitter cold, ice and winds, along with heavy snowfalls, were part of the elements that were endured by the Army Signal Corps and Corps of Engineers.

assigned to the US Army Search and Rescue unit at Harmon were assigned to the job. In three and a half hours the Huskies, assisted by men, hauled the crates to the top of the mountain.

The 21st Signal Corps personnel camped in four railway cars shunted into a siding about eleven miles from the base of the mountain at St. Andrew's. Three of the cars were used for sleeping and the fourth was their mess hall. The men would travel to the mountain base each morning, using gasoline operated rail speeder cars. At times the snow drifted over the tracks three to four feet deep for several hundred yards. The crew had to shovel out the track area in these snow-piled sections. On one particular day, the snow was so deep and the wind so severe, the crews had to turn back to their camp. Unfortunately, they did not hear the approaching freight train with a snow plough, because of the roar of the wind, until it was almost upon them. The speeders were stopped and the men jumped clear, one of the men breaking his leg. The freight train hit the first speeder and completely destroyed it.

Receiving clearance from one of the Newfoundland Railway station signals, speeder crews proceed to their destination. Speeder trailer cars were most often loaded with men and supplies.

There was no highway across the Island of Newfoundland in 1940-41. Temporary roads from Argentia to Colinet through Whitbourne, and Stephenville to Stephenville Crossing were built by the US Army in 1942. Corner Brook, Grand Falls and St. John's had good, but unpaved, roads within their jurisdiction and immediately surrounding areas. The Corps of Engineers constructed roads in Argentia, Pepperrell and Stephenville. There were no roads to Howley, St. Andrew's, Millertown Junction, Gander, Shoal Harbour or Whitbourne. The Newfoundland Railway was the main form of transportation for one repeater station to another. The number of US military personnel continuously travelling by train made up the majority of passengers. The Newfoundland Railway ran three passenger trains a week from St. John's to Port aux Basques and return, taking up to twenty-two hours to complete the 600 mile journey. These trains were known as Number 1 and Number 2 Express. Later, the Americans called it the Newfoundland Express, and as far back as 1943, it was known as the "Newfie Bullet." Two scheduled freight trains made the run each week and the conductor's caboose always carried a few passengers. It wasn't until 1965 that the Trans Canada Highway across the province was completed.

Members of the 21st Signal Corps taking a short break before proceeding on to their destination.

Three-day passes for all the personnel at each of the Repeater Stations were given on an average of once a month. The USO at Corner Brook was the favourite place to enjoy fun and relaxation. Those who were assigned to Corner Brook and Grand Falls were more fortunate. There most of the up-to-date social activities were present, although on a small scale. Each had small clubs, theatres, dance halls, shops and other things that made up town life. The soldiers stationed at Shoal Harbour and Whitbourne would travel to St. John's for their three-day pass activities. The Corps of

American soldiers coming out of church at St. Andrew's and chatting with the local residents in 1942. Often, the only transportation in small Newfoundland communities was the horse and sleigh, or dog team and sled.

Engineers stationed in Colinet would spend their pass at St. John's or back at the Argentia Base.

After May 1945, the end of the European War, activity at each of the Repeater Stations slackened. There was still no direct telephone line communication across the Island, apart from that installed and used by the US Army. After 15 August 1945, when the war with Japan was over, many of the Repeater Stations were reduced to a skeleton crew. More up-to-date equipment installed at Fort Pepperrell, Gander and Stephenville allowed the by-passing of the repeater relay stations. In the beginning of 1947, the Avalon Telephone Company installed a radio telephone system across the Island, and in 1949 leased the US Army lines from Canadian National Telecommunications. CNT acquired the Army Long Line when Newfoundland entered Confederation with Canada. Thus, full telephone communications were installed across and around most of the Island. All the Repeater Stations, except master stations at Harmon and Pepperrell, closed in 1947. Some remote, unmanned communications equipment was installed at some locations. The station buildings were sold in 1949 to private enterprise or demolished. The only original repeater station building remaining is located in Whitbourne, and used as a part of the Volunteer Fire Department there. The cost of the US Government to construct and maintain the communication line, including the Repeater Stations, in Newfoundland from 1941 to 1947 is estimated at $125 million.

When the Atlantic Ferry Command began its operation through Gander in 1942, it was determined that a special radio beacon would have to be set up to guide aircraft towards England, and the return flights to Gander. Wesleyville, Bonavista Bay, was selected as the site of this Direction Finding Station, because of its extreme easterly location and closeness to Gander Airfield. In late 1942 a US Army barge, in tow by a US Army supply boat, left St. John's Harbour, loaded with construction materials and equipment for the proposed radio beacon station in Wesleyville. There were no roads to Wesleyville at the time, so all transportation of supplies and passengers was by sea. The barge and supply boat ran into heavy seas off Cape St. Francis, and the barge sank with all equipment aboard. It wasn't until early 1943 that all materials were in place at Wesleyville. A site known as Alder Ridge, five miles from Wesleyville, was selected, and a road was immediately constructed by the Army Corps of Engineers. One and a half miles from the site was the direction finding station building, with several telephone poles installed all around it. Twenty-four hour guard duty was performed by the soldiers, each taking their turn, along with K-9 guard dogs. Many from Wesleyville will remember the K-9 dog named "Electro," as most people were very much afraid of him. Twenty specially trained technicians from the 136th Army Airways Communications System (AACS) personnel arrived as soon as the facilities were completed and took up their duties. The main objective of the Direction Finding Station was to emit a radio beam (signal) pointing toward Preswick, Scotland by which the Ferry Command pilots could monitor until they were out of range. Radio contact with each aircraft was also maintained until the plane passed the range of transmission. The same radio signals were used whenever an aircraft left Preswick, Scotland, Greenland or Iceland en route to Gander. The last message and signals to be received by a pilot en route to Scotland or England emitted from the Direction Finding and radio beacon at Wesleyville. Conversely, the first radio signal received by pilots on their return flight was the one transmitted by Wesleyville. There was no radar at Wesleyville. The five radar stations operated by the US Army 685th Air Warning Company were located at Sandy Cove, Cape Bonavista, St. Bride's, Torbay and Allan's Island.

Electro, the K-9 war dog that was feared by many during 1943-44.

There was close coordination of radio between Wesleyville, Fogo Island and the United States AACS Signal Corps at Gander. As soon as incoming aircraft got within range of the direction finding station at Gander Airfield, the radio frequency aboard the aircraft

Top left: Sgt. Milton Clark at Wesleyville in 1945, posing with two young ladies, Peggy and Ruby Sainsbury.

Top right: Spring scene at the Ridge, May 1944. The soldiers at times had to shovel up to five hundred feet of snow.

Left: The136th AACS barracks at Wesleyville. The building was built in 1942.

Right: The road constructed in 1942 by the US Army from the Town of Wesleyville to the Ridge where the Direction Finding Station was located.

The barracks' dayroom at Wesleyville, 1945. Note the bed sheet on the wall used as a movie screen. After 13 August 1943, Wesley Hall was approved for public showing of movies, which were well attended by the residents of Wesleyville.

was changed from that of Wesleyville's to Gander's. Wesleyville also reported weather data to Gander every hour of the day.

US Army supply boats, and other government and private marine vessels, were used to supply the 136th AACS Squadron. The station had a jeep, two half-ton trucks and an ambulance. The trucks were used to haul supplies from the Wesleyville dock to the site, as well as transport soldiers into town or bring local residents to the barracks for movies. The ambulance was used to transport both soldiers and local residents to the cottage hospital at Brookfield. The Squadron had a 16mm movie projector and each week the American Red Cross at St. John's would send two movies, plus other film. The soldiers used to invite many from the Wesleyville area to see the movies, some of whom had never seen a movie before. Movies were also shown in the Wesley Church community room.

A large transmitter station was installed at Snelgrove by Windsor Lake, St. John's, in 1943. During the war years this station was manned 24 hours a day; the soldiers from the 21st Signal Corps Squadron at Fort Pepperrell would travel daily by truck to the site, known as "Snelgrove's Hill," overlooking Windsor Lake. US Army personnel from the 3rd Infantry at Fort Pepperrell provided continuous guard duty around the site. There were four buildings on the station: an operations building; warehouse; pump house; and guard duty shack by the main gate. The station closed in 1946.

A high wire security fence surrounded this installation, and only those with special security passes were allowed to enter. No civilian personnel worked here. Windsor Lake, being the water supply for St. John's and Fort Pepperrell, was also patrolled during the war years. The Snelgrove installation was a radio intelligence interception station manned by the 21st Signal Corps. It recorded all enemy submarine signals, as well as all German traffic from North Africa, and was instrumental in the Battle of the Atlantic. The station also intercepted all the radio transmission from the five Radar Stations in Newfoundland. All operations at Snelgrove were "Top Secret."

Sgt. Milton Clark of the 136th AACS at the controls in Wesleyville. Part of the direction finding equipment screen can be seen here.

Max Tiller, a Newfoundland employee at the Wesleyville station. Mr. Tiller worked in the Mess Hall for two years.

One of the vehicles (jeep) assigned to the 136th AACS at Wesleyville, 1943. The driver is Sgt. Milton Clark.

Portugal Cove Road in 1944, by Westcott's Hill, along the shore of Windsor Lake, en route to the Snelgrove transmitter.

Reconstructing the Wesleyville wharf, 1943. Part of the funding was provided by the US government.

8

Goose Air Base, Labrador

> Labrador is important as a possible site for the air base located so as to be of the greatest advantage.

This excerpt is from a report to the War Department in Washington, D.C., by Captain Elliott Roosevelt, USAAC, 21st Reconnaissance Squadron on 6 July 1941. Capt. Roosevelt, the son of President Franklin D. Roosevelt, assigned to the Newfoundland Base Command at Fort Pepperrell, made an aerial and ground survey in June 1941 and selected Goose Bay as the best place for an airfield. The delivery of bombers to Great Britain via Gander Airfield was already in progress by 1941, and it was evident Gander would not be capable of accommodating all the transient aircraft under the RAF and USAF Ferry Commands. Capt. Roosevelt also selected sites for strategic weather stations at Frobisher Bay on Baffin Island, Fort Chimo in Quebec and Padloping Island on the Cumberland Peninsula. These three sites were known during the war years as Crystal I, Crystal II, and Crystal III. A Canadian group from the Dominion Geodetic Society did a similar survey of Labrador in June of 1941 and submitted their report to the Canadian government. The Permanent Joint Board on Defence met immediately after, and within three weeks the Royal Canadian Air Force sent engineers to Goose Bay to design preliminary construction plans for the projected base. In August 1941 the Canadian government awarded a near $10 million contract to McNamara Construction Company of Montreal for the construction of Goose Bay Airfield. The date of completion of the contract was in 1943. However, the runways were completed in six months, as well as some buildings for the Royal Canadian Air Force.

The first of many ships carrying base construction materials to Goose arrived on 17 September 1941. Most of the materials on board were for the first dock at Terrington Basin. As more materials and heavy equipment arrived, work began on the northeast part of the airfield site. The land was cleared in a record six weeks and two snow-rolled runways were quickly placed into service for the winter of 1941-42. The first American plane to land at Goose, on 6 November 1941, was a USAAF mail and freighter aircraft. The very first civilian aircraft to arrive was a ski-equipped Quebec Airways plane on 3 December 1941. It was apparent that when the snow started to melt the two snow-rolled runways would be out of service, so a third runway was started. This one was covered with brush so the frost would not fully penetrate the soil. In the spring of 1942, when the brush was removed, the runway was immediately placed in service. During the spring and summer of 1942 over 623,000 square yards of concrete six inches deep was laid on the two 3,000 feet runways. This, along with temporary buildings, a large concrete gasoline storage tank, warehouses, and a number of intercommunity roads on the base site made Goose Bay operational for the Ferry Command by late 1942.

Goose Air Base, showing two completed 3,000 foot runways, 1942.

The attention of the world turned to the Pacific on 7 December 1941, when the Japanese attacked Pearl Harbor. The United States immediately began to set up the USAAF Eighth Air Force in Britain. The need now was much greater for bombers and fighters to be flown across the Atlantic. Gander was overcrowded with the number of assigned and transient personnel. The United States began to look

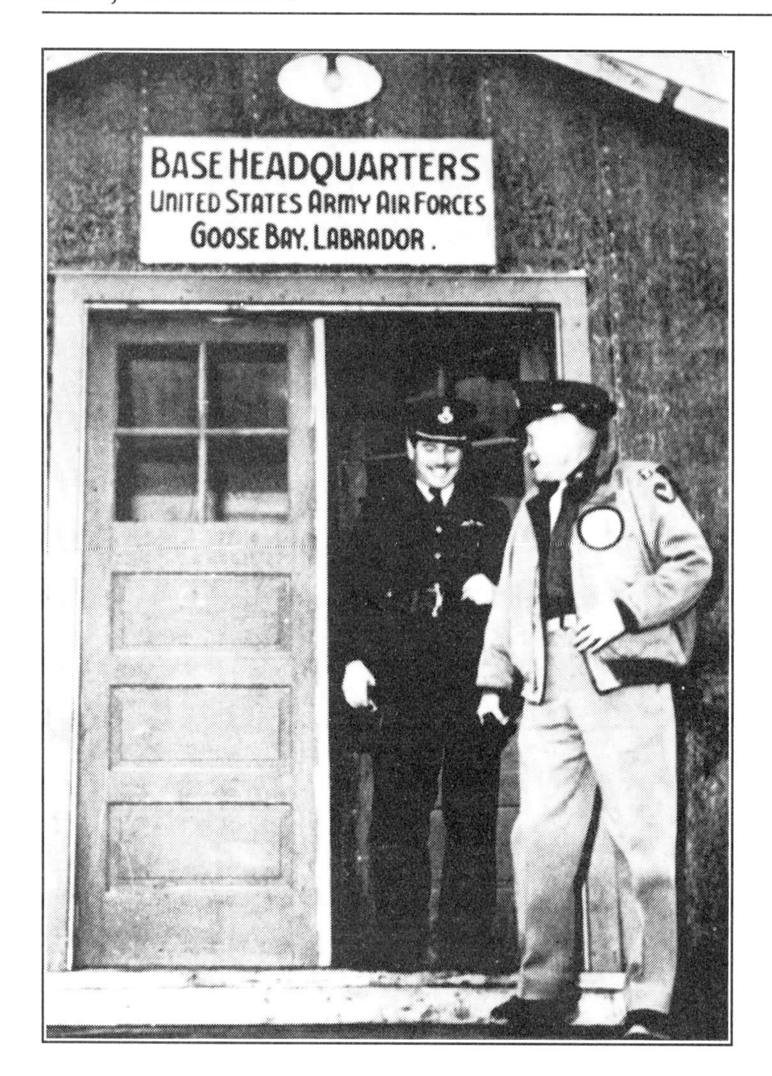

Top left: USAAF Base Headquarters at Goose Bay, 1942. The building was on the RCAF side of the Base, as no American facilities were constructed up to that time. (PA-154002)

Top right: This photo was set up as a gag to depict the heavy snowfalls in Goose Bay.

Bottom: The USAAF Base Headquarters on the American side of the airfield, 1943. This was one of the typical type of winters in Labrador.

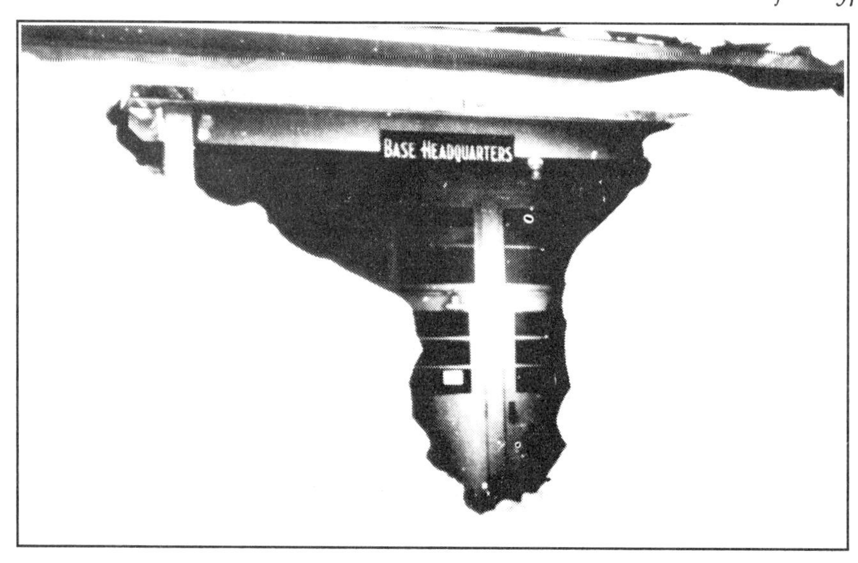

upon Goose Bay as the answer to its requirement for an alternate airfield and space to perform. By 7 December 1941 the RCAF already had a strength of over 3,000 personnel at Goose. An agreement was reached with the Permanent Joint Defence Board to allow the United States to build a base at Goose Bay on the southeast side of the airfield.

On 31 January 1942 the United States government signed a contract with Northeast Airlines. The airline would provide maintenance for service between Presque Isle, Maine and the bases which were being built in Newfoundland and Labrador. The United States government was to provide the aircraft for transporting personnel and materials for the air bases under construction. It was not until March 1942 that the first USAAF officer arrived at Goose Bay to plan for the construction of a weather station. The first contingent of enlisted personnel did not arrive until April of that year. Goose Bay, like Gander, was shrouded in secrecy. The words Goose Bay were seldom used in correspondence, on printed information or in local news sheets.

The following extract of the early history of the United States military activity at Goose Bay is quoted from "A Brief History of the 95th Strategic Wing and Goose Air Base," 15 February 1967, by S/Sgt. Maurice A. Miller, Jr., USAF historian, who was stationed at Goose Bay.

> Initially, the USAAF cadre was assigned to the Eighth Army Airways Communications Detachment which maintained its headquarters at St. John's, Newfoundland. By 21 April 1942, less than three weeks after arriving, the personnel had the base radio station in operation. Then, by 5 June, there were seven officers and 21 enlisted personnel at Goose with 24 enlisted men attached. The 1st Provisional Ferrying Squadron of the North Atlantic Ferrying Command was assigned to Goose on 5 June 1942. Although Goose was not in any position to initiate ferrying service at that time, RCAF and USAAF personnel met to plan for the eventual handling of large aircraft movements between the United States and the United Kingdom....
>
> During July 1942, the USAAF Ferry Command was absorbed into the Air Transport Command. Goose Bay continued operating under the North Atlantic Wing and on 4 August 1942, the 29th Ferrying Squadron was activated at the station. By the end of 1942 there were 5,000 American, Canadian and British military on the base and over 3,000 construction workers.
>
> The first movement of tactical aircraft between the United States and the United Kingdom, staging through Goose, began in June 1942. During this month Goose initiated ferrying procedures. The first operation that the base participated in had the code name of "Bolero." This was the build-up of men and material in England for the eventual invasion of Europe. In this

first month 62 tactical aircraft deployed to England through Goose....All in all, 662 tactical aircraft deployed through Goose to build up the Eighth Air Force during the last half of 1942. This total only reflects American USAAF aircraft. The RCAF on the Canadian side of the base handled all those planes being flown to England by the RAF Ferry Command.

...In December 1942 Goose played an important role in the CRIMSON project. Northern bases were developed to aid in ferrying and refuelling aircraft deploying to Europe. Eventually, Bluie West One and Bluie West Eight were constructed as part of the Northern Ferry Service. These bases, in Greenland, eventually became Narsarssuak and Sondrestrom.

...By January 1944 the Allied forces were preparing for D-Day; Goose continued servicing aircraft deploying through Europe. There were 5,262 US aircraft staging through the base during 1944 with the greatest workload of the year occurring 4 July when 149 bombers landed. Goose handled more cargo and passengers, both east and westbound, than any other overseas station in the North Atlantic Wing. During the period 1941-45 it is estimated the airport handled over 25,000 US, Canadian and British aircraft....

The USAAF Service Club, 1944. This was a recreational facility for all the American military.

Over 240,000 military personnel returned to North America via Goose, Gander and Stephenville without a single fatality. It was the largest recorded military airlift of personnel.

Goose Bay was bursting at the seams. All barracks had double bunks in them, and in some instances, many sections of buildings had to be temporarily converted to accommodations. There were two civilian barracks on the base, and all others lived at Happy Valley, Mud Lake, Otter Creek, North West River and in tents set up along the base perimeter. There was one area set up on the base, known as "E" area, where several male, temporarily employed civilians were quartered. Temporary tent-type accommodations were set up at "G" and "H" areas to house in-flight crews overnight. There were times when the total population of Goose Bay far exceeded 9,000 people. The only women on the base, up to 1944, were the Red Cross girl, two nurses on the American side and one female secretary on the Canadian side. The villages of Happy Valley and Mud Lake were "OFF LIMITS" to all military personnel. In April 1945 a squadron of Women's Auxiliary Corps (WACS) arrived at Goose. One of the enlisted men's barracks was renovated and became the WACS barracks. Later in 1945, forty-five Canadian girls were recruited at Moncton, New Brunswick, and flown to Goose. They were also housed in the WACS barracks. These girls worked in the post exchange, offices and NCO's Club, telephone exchange and at the Hotel DeQuink, which was an officer's transient barracks.

Duty at Goose Bay was basically complete isolation: no outside base roads; no adjoining towns (except Happy Valley and Mud Lake); and nowhere to go on a three-day pass except to enjoy some

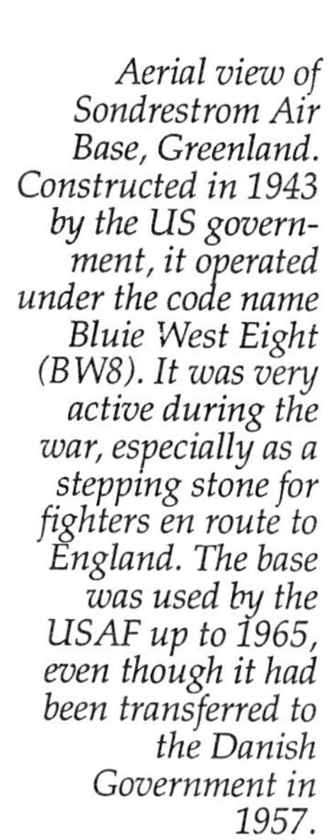

Aerial view of Sondrestrom Air Base, Greenland. Constructed in 1943 by the US government, it operated under the code name Bluie West Eight (BW8). It was very active during the war, especially as a stepping stone for fighters en route to England. The base was used by the USAF up to 1965, even though it had been transferred to the Danish Government in 1957.

Narsarssuak USAF Air Base, Greenland. This air base, known during the war as Bluie West 1 (BW1), was quite active. Used as a stopover for bombers and fighter aircraft en route to England, it handled thousands of planes. The US closed the base in 1957.

skiing, fishing and hunting. On 11 December 1943, the RCAF started Radio Station VOUG, and through special radio receivers they were able to tie into the American network and broadcast these programs locally. New movies were shown at the base theatre every two days. The United Service Organization (USO) contracted more USO shows through Goose than any other installation. Big time stars such as Bob Hope, Frank Sinatra, Sammy Davis, Jr., Lana Turner, Andy Devine, Francis Langford, and many, many more performed at the Base Theatre at Goose. Both civilians and military attended these functions. The base service club, gymnasium and the Officers' and NCO's mess were usually filled to capacity each evening. The tour of duty during the war years was for the duration. After 1945 it was eighteen months before rotation to the States or to another base. The workload placed on everyone kept them all busy with little time to brood over the nightlife back home.

In accordance with the agreement on the disposition of Goose, the RCAF assumed control of the base in February 1946. The US forces deployed from Europe through Goose. During this time a possible first in aviation history, not involved with the redeployment, occurred in Labrador. A new invention called a helicopter rescued crewmen that had survived a PBY crash off the Labrador coast.

After the end of hostilities, certain administrative changes affected Goose. Brigadier General Caleb V. Haynes, commanding general of the Newfoundland Base Command, assumed control of all bases in Labrador. This first unification of the northeast bases occurred 31 March 1946. It also formed the basic complex for the activation of the Northeast Air Command (NEAC) within the next four years.

Jets from the 59th Fighter Interceptor Squadron undergoing daily maintenance and checking on the parking apron.

RCAF Station, Goose Bay, winter 1943.

Civilian workers lining up by the mess hall for a mid-day meal, 1949.

The civilian barracks area in the early days of Goose Bay.

Between 1947 and 1948 Goose was subjected to routine operations normal for a refuelling stop. Yet many experiments were conducted in search and rescue. At that time a weather station being built at Indian House Lake needed extensive lumber supplies. With the requirements levied to the Air Force during the spring months, techniques for dropping lumber and other supplies were attempted. After many drops it was found that lumber bound in lots of 200 pounds could be dropped into the lake without a parachute by a low-flying aircraft at minimum speed. This technique was soon developed so that eggs could be dropped into the lake without any damage.

Also in 1948, sixteen F-80 Shooting Stars of the 59th Fighter Group stopped over at Goose en route to London, nonstop. This was the second successful trans-atlantic nonstop flight for jets. The first was a few weeks earlier when six RAF Vampire jets took off from Goose to London.

Although most of the Zone of Interior (ZI) airfields were on a stand-by basis, Goose was still active. In 1948 the mission of the base was to provide maintenance of stopover aircraft, to train permanent party personnel, and to furnish logistical support for the three weather stations at the Crystal sites and the detachments at Indian House Lake and Cape Harrison.

In January 1949 a further step was made towards the unification of the old northeast bases. A direct communications system was established between Goose and Pepperrell AB, Newfoundland. Previously, any communication between these stations had to be sent through Ernest Harmon AFB. This action was accomplished at the right moment for, in October 1950, the old Newfoundland Base Command was inactivated with the Northeast Air Command (NEAC) being activated at Pepperrell. Along with the activation of NEAC, the 6603rd Air Base Wing was activated at Goose and the installation became an integral part of the regional defence system. By this time the Strategic Air Command was being supported by Goose.

...Another reorganizational change occurred during 1952. Headquarters NEAC redesignated the 6603rd Air Base Wing as an air base group commensurate with other such organizations throughout the command. During this time Goose actively supported SAC operations, and in July 1952 initiated project SEAWEED. The first jet aircraft assigned to Goose were F-94B, of the 95th Fighter Interceptor Squadron, on 3 November 1952. The 59th Fighter Interceptor Squadron was activated at Goose in January 1953 to provide fighter protection of this base. This unit was originally on an indefinite duty tour, but this eventually lasted until January 1967 when the 27th FIS assumed the duties of the 59th FIS. Besides this, Goose also received control of Frobisher Bay from NEAC. This occurred 12 June 1953. In October a companion unit from the 59th FIS was activated at Goose: the 22nd Helicopter Squadron was assigned H-19A and H-19B helicopters.

During the first half of 1954 the construction program curtailed because of the Korean War, again commenced at Goose. The airfield was repaved. Nine aircraft hangars, four

Top: Runway inspection of the 59th Fighter Squadron pilots and jet aircraft.

Centre: F-94B jet fighters lined up on the ramp.

Bottom: Bowling alley at the recreational hall,Goose Bay, early '50s.

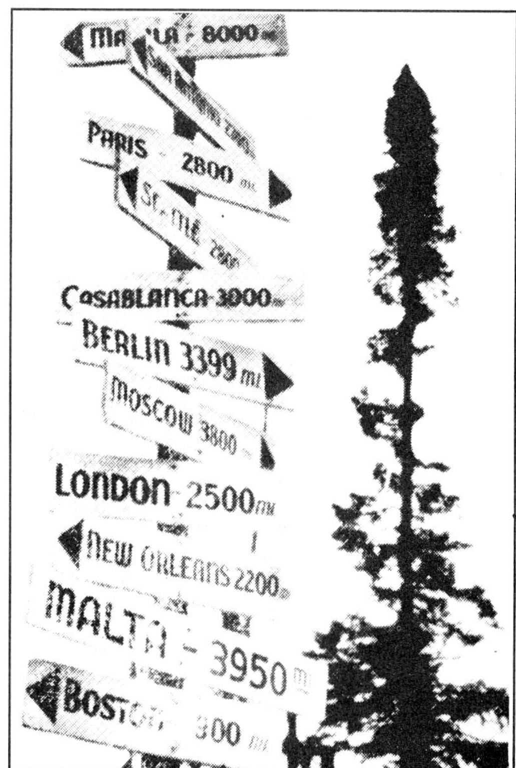

Top: Winter scene of the old base chapel, Goose Air Force Base.

Bottom: Directional and mileage sign erected by a group of USAAF personnel at Goose Bay, 1945.

concrete warehouses, the steam power and heating plant, navigation-aid facilities, two fire stations, a heated motor vehicle garage, and several shop buildings were constructed. Work began on the chapel, theatre, PX, new base operations building, and headquarters building.

Goose was again affected by reorganization when NEAC deactivated the 6603rd Air Base Group and organized the 6606th Air Base Wing. The base was not only required to support aircraft, but to participate in scientific research and exploration projects. The base was called upon to give local support to scientific parties in Labrador for recording the solar eclipse of 30 June 1954. The support given by the base aided immeasurably to a better understanding of eclipses and the sun itself....

While flying over the northeastern part of Labrador, a pilot spotted a bright green lake that was completely round in shape. He felt this was rather unusual since the other Labrador lakes had a blue shading to them. In July 1954 Goose supported a scientific expedition trying to locate this lake to determine the cause of the phenomenon. When the party arrived at the lake, they found that it was a deep, bright green. It was perfectly round, and absolutely void of any form of life. Several explanations were offered toward the solution of this mystery. The first theorized that the hole was caused by a meteor, hence its name of "Crater Lake." It was this group's contention that a meteor had probably struck the earth before the ice age and that all evidence of the meteor had been destroyed. Another theory advanced by another group stated that the lake had been caused by glacial or ice movements in the ice age. The scientists finally took water samples from various depths and returned to their respective laboratories to attempt an explanation on the origin of the lake and why there is no life in it. Today there is still no answer, just the same questions remain.

...In 1956 several changes occurred that affected the organizational structure of Goose again. During the first half of the year Goose received District 3, Office of Special Investigations, and the 5th Weather Group's Detachment 4. Toward the end of December plans were announced for the reorganization of the northeast area. Subsequent changes gave Goose the added responsibility to support the 59th FIS, the 920th, 922nd, 923rd, 924th, and 936th Aircraft Control and Warning (AC&W) sites, in addition to certain early warning gap-filler stations along the Labrador and Newfoundland coasts. Goose was also required to support Detachments 4 and 7 of the 6th Air Postal Squadron and detachments of the radio relay squadrons.

Headquarters Strategic Air Command (SAC) designated Goose a Category One installation on 9 January 1957 with the base giving top priority support to SAC operations....The Northeast Air Command, in existence for seven years, was deactivated 1 April 1957. All its bases were divided between SAC and ADC. Then, on this date, the 6606th Air Base Wing was deactivated and a strategic wing, destined to outlive any other unit at Goose, was formed; this was the 4082nd Strategic Wing

Goose Bay Air Force Base, 1957-58. In the centre foreground is the Goose Hilton, the six storey dormitory. While the photo only shows part of the American base, a portion of the RCAF side can be seen at the top.

Goose Air Force Base, 1953, showing the complete American and Canadian sides of the airfield. The dock at Terrington Basin can be seen in the upper centre of the photo.

that would live through the end of the propeller-driven alert aircraft and into the jet age.

At the time that SAC took over Goose, there were 3,300 military personnel and over 700 civilians working on the American side of the base. The KC-97 refuelling tanker aircraft started to use Goose for many of the Squadron's operations. The new B-47 Bomber began its early flights from Goose to overseas destinations. The flight crew participated in many projects to determine the capability of both aircraft and aircrews to function under severe climatic extremes....Work was completed on the new dental clinic, commissary, dependents' elementary school and an NCO open mess. Construction began on a new civilian dining hall, airman's dining hall (Herbert Hall), a new electrical power plant and heating plant. The "Goose Hilton," the highest building in Labrador, and dormitory for 500 men, was completed and occupied in November 1959. The year ended with an award of a contract for new alert facilities for the Fighter Interceptor jet crews.

The history of Goose indicates that after the War years, the base alternated among periods of relative calm, reorganization and construction programs. By November 1958 the USAF relinquished control of Frobisher Bay to the RCAF. A new working agreement was signed in June 1958 with the Canadian Department of Transport handling the logistical requirements formerly furnished by the USAF. Airdrome improvements resulted in safer environments for the refuelling operators of the KC-97 propeller-driven aircraft. Projects were completed to light the ramp for taxiing and manoeuvring of large numbers of aircraft at night. Seeding and planting of grass to prevent the severe dust storms during the summer was a major undertaking....Goose turned green in the summer of 1959. On 1 January 1959 the Strategic Air Command at Goose came under

The original dock site at Terrington Basin at Goose Bay. Note the RCAF buildings and part of the runway in the centre.

the jurisdiction of the 45th Air Division at Loring AFB in Maine. In May 1959 the exclusive "Icebreaker" Service Club was completed. On 12 April 1960 the first KC-135 jet tactical tanker arrived at Goose. This unit was followed by many more that were stationed on the base for over six months. In 1960-61 the 54th Air Rescue Squadron was activated at Goose Bay. At the close of the shipping season all the remaining US Army Transportation Unit members withdrew from operating the Port of Goose Bay. The Army Unit was first assigned back in the early fifties. Both the USAF and the Canadian Department of Transport [DOT] came to a working agreement for the operation of the port facilities. DOT agreed to handle the continuity of civilian personnel to unload ships while the USAF furnished certain support services. In October 1962 the USAF began a $350,000 project of redredging the two channels which allowed ships of 35 feet draught to enter Goose Bay.

A more significant organizational change took place on 1 July 1960 when the KC-97 Goose Tanker Force was assigned to Sondrestrom Air Force Base in Greenland. Several KC-97 refuelling tankers were placed on alert there. The object of stationing the KC-97's there was that they did not have the range of the new KC-135 jet tankers.

...In October 1965 Goose Bay went down in history once again when the 4082 Strategic Wing supported the historic Operation "Big Lift," in which an entire US Army division and its support units were deployed to Central Europe. Goose supported 19 C-133 and 48 C-124 aircraft staging through the base on 22 October 1963. In 1966 when Ernest Harmon Air Force Base closed, Goose AFB assumed many of the functions previously performed by Harmon. Towards the end of 1966, the 4082 Strategic Wing was deactivated and it became the 95th Strategic Wing. Many Air Force bases in the United States and overseas closed in 1966, however, the strategic importance of Goose Air Base was still top on the list in Washington, D.C. The Fighter Interceptor Squadron was transferred to a new base in the States. New, modern, large aircraft were assigned to Goose, Like the B-58, C-07, HU-16 and U1A. Although many USAF units were assigned to Goose Air Base over the years, change in organization, mission and support were the keys to the history of the base.

The PX at Goose Bay Air Base, 1950s.

Goose Air Base continued to prosper for the next ten years. The RCAF had a small number of personnel on their side of the base and, on 1 August 1967, they turned over all their operations to the Canadian Department of Transport. Some RCAF personnel remained to operate Radar and Traffic Control, as under the original US/Canadian agreement covering the USAF presence in Goose, this function of the base had to come under Canadian Military command. In 1971 the Canadian Forces Base on the north side of Goose phased out. All buildings were turned over to the Province of Newfoundland. The remaining RCAF and RAF moved over to the American side. A small contingent of RCAF personnel took over the operation control on 12 July

1971 of the NORAD Control Centre and the Long Range Radar Station at Melville, Goose Bay.

In 1973 the 20-year lease, signed in 1953, was extended for six months. In July of that year the land, buildings and improvements of the USAF were turned over to Transport Canada, and a new agreement was signed to provide services and facilities to the USAF until 1 July 1976. On that date the Canadian Department of Public Works assumed responsibility from the Canadian Department of Transport for maintaining roads, buildings, operating the heating plant and providing snow removal, water service, fire-fighting, and other miscellaneous services. The DOT continued to operate the airport.

USAF Headquarters in Washington, D.C. announced the deactivation of the 95th Strategic Wing at Goose Bay in September 1976 and that their active involvement in the use of Goose Air Base would cease. The final closing was on 1 October 1976. A few USAF personnel have remained at Goose since that time to look after the interests of any U.S. military aircraft stopping over for service....The net worth of the American side of the base, including land improvements over the years, is estimated at over 250 million dollars.

USAF search and rescue dog farm, Goose Bay Air Force Base, 1955. Rescue dogs were used at Goose as far back as 1942 for search and rescue of aircraft crews and others. The dogs were trained to parachute into the wilderness area where downed air crews were known to be located. Many a downed pilot and crew member owe their lives to these highly trained animals.

Today, Goose Bay Airport is once again rejuvenated. The Canadian government, along with the NATO nations, are trying to have the airfield designated as a NATO training centre for low-flying jet aircraft. Already the British Royal Air Force, Royal Netherlands Air Force, West German Air Force and the United States Air Force are actively engaged in these exercises. During the year 1989, over 6,000 flights moved through the air base.

The Town of Happy Valley, incorporated in 1955, is a flourishing community. The main industry is catering to the needs of the air base with numerous services. Commercial daily flights move in and out of Goose. Many new and attractive buildings have been constructed at Happy Valley in addition to those which were turned over to the town by the USAF. The original Town Hall at Happy Valley was once the old base chapel. Several buildings were sold to private enterprise. The population of Happy Valley in 1981 was 7,103, a long cry from the scattered families living there in the 1941-46 era. The majority of the people there work on the base, or in support activities for the operations of the airport facility.

Townsite of Happy Valley, Labrador, 1947.

Thule AFB, Greenland, showing the AC&W Station on "P" Mountain. Thule was one of the most expensive air bases in the world.

AC&W Station at Jerry's Nose, Port au Port, thirteen miles from Harmon Air Force Base. This station was built on fifty acres and had thirty-two buildings on the site and was manned by over one hundred personnel. The elevation was 1,250 feet above sea level.

9

Radar and Aircraft Control and Warning Stations

Radar was first used by Great Britain in late 1939 to detect approaching aircraft of the German *Luftwaffe*, as they neared the shores of England. The British radar network was superb in its day and was underestimated by the Germans until 1940. On 13 August 1940 the German High Command ordered an extensive air raid on British radar sites on the southern shores of England. Germany had radar as early as the British, however they did not fully evaluate its effectiveness. The United States, in its effort to strengthen its defence in strategic areas outside the continental United States, installed radar in many locations. One of them was Pearl Harbor, where the early warning devices were installed in late 1940. Minutes before the Japanese attack on Pearl Harbor on 7 December 1941, US radar picked up the approaching Japanese planes approximately 150 miles away. The planes were launched from a Japanese aircraft carrier 200 miles from Pearl Harbor. Although the warning was given by radar operators at Pearl Harbor, it was practically ignored by the US Navy Command on that fateful Sunday morning. After the 7 December attack, most US Navy ships were ordered to install radar units. The USS *Pollux* supply ship which ran aground at Lawn Point on 18 February 1942 was carrying radar units to Argentia, to be installed on US ships operating out of the US Naval Base. Radar was not used on aircraft as yet, because the complicated equipment required for planes was still in the experimental stages.

Realizing the strategic importance of defending Newfoundland from enemy attack by air, American and Canadian anti-aircraft batteries were set up at many points along the Eastern Seaboard, especially on the Avalon Peninsula, in Placentia Bay and on the west coast around Stephenville. The Canadian Forces had full defence responsibility for the area around Gander and Goose Bay and the St. John's Harbour. The United States knew, however, that this wasn't enough, as no radar devices were installed to detect approaching enemy aircraft and to provide an early warning. In March 1942 the US Army Signal Corps 685th Air Warning Company was assigned to the Newfoundland Base Command, with their headquarters at Fort Pepperrell. The initial contingent of personnel of the 685th was only about forty trained personnel. Their mission in Newfoundland was Top Secret. During the spring of 1942 it was decided that the use of radar in Newfoundland was an absolute necessity, especially with the Atlantic Ferry Command in full operation at Gander, Goose Bay and Harmon Field, and the US Naval Base activities at Argentia. The initial site chosen was Sandy Cove, Fogo Island. The very first US Ground Radar Early Warning Station in the North Atlantic area was to be installed. None of the people at Sandy Cove, or on Fogo Island, knew what was going on when the first 685th Air Warning Squadron crew arrived at Sandy Cove in the early summer of 1942. Work began immediately on the acquisition of land and the construction of several temporary buildings required for the station. All supplies had to be shipped by sea transport, using US Army supply boats operating out of St. John's. By the end of the summer of 1942 the station was ready for the installation of equipment and manning. Fifty-two personnel from the 685th Air Warning Squadron at Fort Pepperrell were sent to Sandy Cove as the permanent operating unit. The detachment consisted of three officers, one doctor, two medical attendants and forty-six technicians.

The radar shack was located one mile from the main barracks area. It was secured by a barbed wire fence and continuously patrolled by soldiers and K-9 dogs. No one without Top Secret clearance was allowed near the area. The residents of all communities where the radar units were installed had no idea, until late in 1944, what the Americans were doing there, except that they were engaged in weather observations. There were seven 50 calibre machine guns within sandbag-enclosed installations, mounted in strategic positions around the radar site. Altogether there were ten buildings on the site, including thc restricted radar and radio equipment on "Radar Hill." It was self-contained with its own power plant of three 30 kw Cummings diesel generators, a station telephone system, well water and treatment system, waste disposal plant and fuel storage tanks. Besides this, a headquarters building, barracks, mess hall, recreation building and warehouse structures made up the complement of the station.

The radar unit was called Special Control Radar (SCR) 270, and the antenna was rotated by two 3 hp motors on the tower. The antenna was 80 feet high and 30 feet wide. Each motor had a 30 pound counter weight. This was the same type unit installed in 1941 at Pearl Harbor. It had a 150 mile range, not much compared to radar units of today's standard, however it did provide an early warning and fast notification to alert Gander and the Newfoundland Base

Seven 30-calibre machine guns were strategically located around the Fogo Island radar station, in sandbag enclosures, and used to protect the station against enemy attack. The station also had several K-9 dogs, which were used to patrol the perimeter of the station.

Top right: The main part of the radar complex, showing the barracks, mess hall and operations building. The radar antenna can be seen in the background on Radar Hill.

Centre right: The men of the 685th Air Warning Squadron at Sandy Cove, Fogo Island, July 1943. Altogether, fifty-two men manned the station.

Bottom right: The US supply boat Captain Mitchell *made regular runs from St. John's to bring supplies, men and equipment to Sandy Cove from 1943 to 1944.*

Headquarters at Fort Pepperrell. Gander had the 432nd Technical Bomber Squadron with fighters assigned to the base, and they were constantly on the alert for warnings concerning an enemy or unidentified aircraft. The radar station at Sandy Cove was the most active of the five that were installed on the Island of Newfoundland. It tracked all the traffic from Gander to Goose and out into the Atlantic. Communication to Gander and St. John's was by radio with two transmitters, a 600-watt main radio and a 150-watt auxiliary.

In the fall of 1942 the installation of four other radar units was begun by the 685th Air Warning Company. The locations were at Cape Bonavista, Torbay, St. Bride's and Allan's Island, close to Lamaline. By early winter of 1943, all were completed. The US Army supply boats, the *Captain Mitchell* and *T-4*, brought all the supplies to remote areas and continued to supply all stations, except Torbay. Again, fifty-two members of the 685th AWC were assigned. All of these personnel were fully trained radar technicians, selected from Maine to Florida. S/Sgt. Joe Santomas, from Hammonton, New Jersey, was the technician who installed and supervised all the maintenance on the radar units. Santomas, now retired from the US Army, continuously travelled by the US Army supply boats to each location. He maintains that in 1943 all the radar antennae were changed and these gave each station a range of 300 miles. All radar stations were in constant contact with one another by a separate radar network. All transmissions to St. John's were directed to the Top Secret receiver/transmitter at Snelgrove, near Windsor Lake. The secret code names for all the radar sites were: Torbay *Prime*; Cape Bonavista *Second*; St. Bride's *Trio*; Sandy Cove *Quad*; and Allan's Island *Cinco*. All radar information was sent to a plotting centre at headquarters, Fort Pepperrell, to track the aircraft coming and going. The plotting centre also sent messages to the radar units concerning lost aircraft or those needing navigational assistance. Both American and Canadian personnel manned the plotting centre.

The transmitter in the Communications Room on Radar Hill, 1943. The unit operated on 600 watts, 1900 KC at night, and 3800 KC during the day. During its two years of operation, not one moment of down time was experienced. S/Sgt. Joe Santomas is at the controls.

By 1942, Argentia had its own ground radar unit installed in the operations building, as well as on all the patrol boats and ships in Placentia Bay and on the Atlantic. The need for radar at Harmon Field was not considered a necessity as any approach of enemy aircraft would have to come from the east. The RCAF installed and operated radar at Goose Bay late in 1942, as well as in Gander. When, in late 1944, the Allied Forces in Europe were obviously turning around the war, the personnel strength at each radar location was gradually reduced. Furthermore, the installation of radar in 1943 on all Allied patrol aircraft, as well as bombers and transports, provided a tight network of early warning for the whole North American area, and the American and Canadian seaboard. Many of the technicians were transferred to the Pacific area of hostilities. All the radar sites closed down in early 1945 and buildings were turned over to the Government of Newfoundland. Today, there is little or no evidence of their existence, except in the memories of some of the local residents.

The advent of the Cold War with Russia in the early 1950s, and the need to provide "a necessary air umbrella" over the northern hemisphere, prompted the North American Air Defence Command (NORAD) in Colorado Springs, Colorado, to select sites and build a number of Aircraft Control and Warning Stations from Greenland to the Arctic and along the Eastern Seaboard. The project was a joint US-Canadian operation with the United States concentrating on the Eastern Seaboard, especially in the Newfoundland and Labrador area. Once again the strategic location of Newfoundland was recognized.

Construction began in 1951. Each major area selected was supposed to be a self-contained, self-supporting unit that would operate on a continuous basis in all areas, some of which were very remote. The first installations were built at Thule Air Force Base in Greenland, Goose Bay, Labrador, Jerry's Nose on the Port au Port Peninsula, Gander Airport, Red Cliff outside St. John's, Argentia, Frobisher Bay, Sagalek, Resolution Island, St. Anthony and Cartwright in Labrador. To maintain alertness and interception of unknown aircraft or ballistic guided missiles, squadrons of jet fighter aircraft were stationed at Ernest Harmon AFB and Goose Bay AFB. F-102A supersonic Delta Dart fighters of the 323rd Fighter Squadron were the first to be stationed at Ernest Harmon. In June of 1960 these aircraft returned to the United States and a new squadron of F-89 subsonic Saber Jets was assigned to Goose Bay, Labrador.

When the 64th Air Division took over the Northeast Air Command in 1960 they determined that there were gaps in the warning system. Therefore, smaller AC&W sites, "Gap Fillers," were built along the Labrador coast and in Newfoundland. Elliston, Bonavista Bay and La Scie, White Bay were constructed in Newfoundland and five locations in Labrador, namely Hopedale, Makkovik, Cut Throat, Spotted Island and Fox Harbour.

A separate communication link was required between Thule AFB in Greenland and NORAD Headquarters in Colorado Springs. A 1,200 mile underwater cable from Cape Dyer on the east coast of Baffin Island was laid to Hampden on the west coast of Newfoundland and from there, overland to Deer Lake and Wild Cove, and undersea again to Cap de Roches on the Quebec shore. The Canadian Overseas Telecommunications Corporation was in charge of the project. They contributed $11 million to the project and the United States covered $8,410,000 of the cost between Baffin Island and Newfoundland, and the cost from Thule Air Force Base to Cape Dyer.

All of the AC&W sites were known as the "Pine Tree Line." Pine Tree was a back-up for the Arctic Dewline Project, which was built

Resolution Island AC&W Station in Resolute Bay. Aircraft would land on the lake in summer or winter. Support buildings for the site operators can be seen on the lower road. The ride up the side of this mountain was an experience in winter!

St. Anthony AC&W Station opened in 1956 and closed in 1968. A road from the site to the Town of St. Anthony was provided and maintained by the radar station. Stations at Cartwright and Hopedale in Labrador closed at the same time as did St. Anthony.

The radar dish at Sagalek was the same as that installed at other AC&W stations. This winter scene shows the difficulties of maintenance in sub-zero weather.

after the war years as an early warning system of attack on North America. Sites for all the AC&W stations had to be on the highest altitude possible in the given location. This was to eliminate any type of surface interference with the delicate and sophisticated radar equipment. Each station had a fair sized acreage and all permanent buildings had to be constructed. Those in the Far North required the capabilities to operate in more than extreme cold conditions. Because of their elevation, each station was subject to severe winds, excessive snow accumulation and extreme low temperatures. It was necessary to include such structures as dormitories, shops, warehouses, dining halls, recreational buildings, as well as the operations centre. Some had to have interconnecting tunnels or above-ground interconnections between buildings. Although most of the large stations had an Air Force Base nearby for logistical support, each facility was more or less on its own.

In the early 1950s a special Aircraft Control and Warning Control Centre was constructed on the White Hills at Pepperrell. Built entirely of 8 to 10-inch concrete, the building, number 1050, was named the "Air War Room." The structure remains today, with some renovations and outside vinyl siding, as the Department of Culture, Recreation and Youth of the Newfoundland government.

The Headquarters, Northeast Air Command, and later the 64th Air Division, were instantaneously notified of any approaching unidentified aircraft or object, at about the same time notification was received by NORAD at Colorado Springs. It was a most unique network of defence. Lone jet fighter interceptors were continually in the air, while reconnaissance and surveillance type aircraft covered 14-hour patrol.

Pepperrell served as the control centre of the "Eyes of the North—the Sixty-Fourth," the 64th Air Division (Defence) whose mission was to maintain a constant radar watch of the northeastern approaches to the continent and to intercept and identify any unknown aircraft within these approaches and, if necessary, destroy aggressors.

The 64th had a three-fold responsibility: to the North American Air Defence Command (NORAD); to the Continental Air Defence Command (CONAD); to the US Air Force's Air Defence Command (ADC). The division's responsibility to NORAD was an operational one in that the work of actually guarding the North American continent comes under NORAD. The division accomplished this responsibility through the Northern NORAD Region, the Canadian Operational Headquarters at St. Hubert, Quebec. Because neither Denmark nor its bases in Greenland was a part of NORAD, the 64th Air Division units assigned there operated under the Continental Air Defence Command. CONAD was strictly a United States Command, not a bi-national one as NORAD was. The Air Defence Command responsibility was one of training and logistics. The ADC equipped, administered and trained the men of the 64th for operational use.

The New York Air National Guard Unit, the 152nd Aircraft Control and Warning Group, was called to active duty at White Plains, New York, in August 1951. In March 1952 it was designated the 64th Air Division (Defence) and advance parties were sent to the northeast where they immediately began to set up Aircraft Control and Warning Stations. On 3 November 1952 the first aircraft, F-94B's of the 59th Fighter Interceptor Squadron, arrived at Goose Air Base, Labrador. A detachment of four of the squadron's 94s was immediately sent north to Thule Air Base, Greenland.

From these modest beginnings the division grew to a 2,000 mile chain stretching from northern Greenland down the eastern Canadian coastline to Newfoundland; an area of responsibility equal to the size of the United States. The air division was composed of air defence groups, fighter interceptor squadrons, AC&W squadrons, and intermediate surveillance stations. After the deactivation of the Northeast Air Command in 1957, the division took over Pepperrell Air Force Base and its units. The vast distances between the units of command tended to group them into three different "island" areas. These areas had their centres at Thule Air Base, Goose Air Base, and

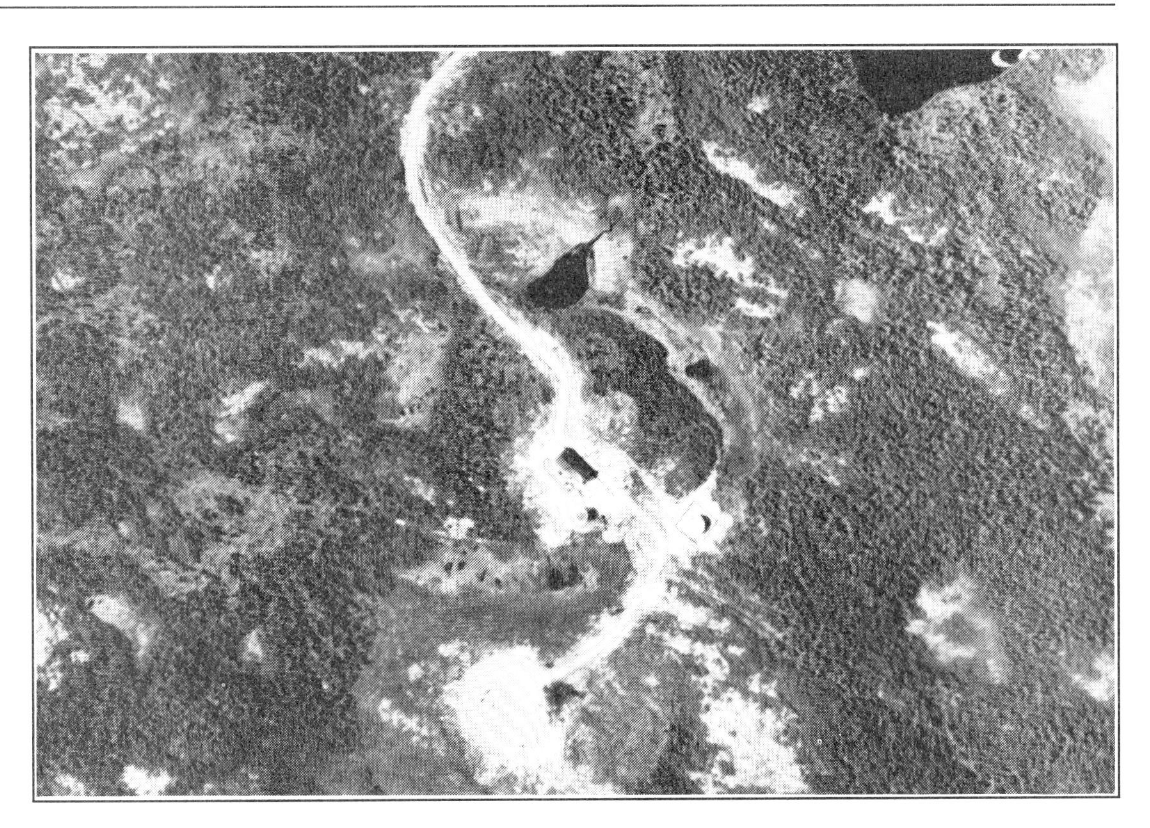

Above: La Scie, White Bay, Gap Filler AC&W Station. As noted, some stations were much smaller than others. This station closed in 1960.

Right: The radar station at Elliston Ridge, Bonavista Bay. This AC&W site was larger than some Gap Fillers since its mission included additional functions. This station closed in 1960.

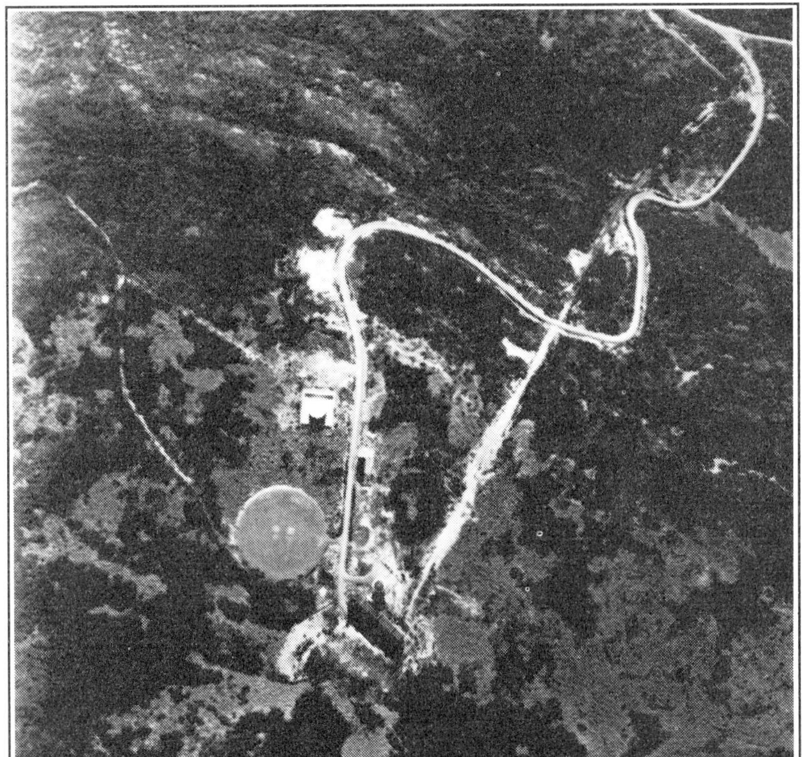

the Island of Newfoundland. Each of these complexes contained a fighter interceptor squadron and several aircraft control and warning squadrons.

The fighter units of the division operated under conditions unlike those encountered by any comparable organization in the United States Air Force. The intense cold and severe arctic winds combined to hamper operations and maintenance. Magnetic variables in this area made compasses unreliable, thus navigational problems were complex. Pilots of the northernmost fighter unit, the 327th, patrolled the arctic icecap through the summer when the sun never set and through the winter when the sun never rose.

In the execution of the mission, radar operators in the AC&W squadrons tracked the movements of aircraft sighted and passed speed, height and direction information to direction centres and fighter interceptor squadrons. Fighter aircraft were airborne in minutes after an alarm was given, and closed in on unidentified planes by means of radar detection equipment in the rocket-armed Northrop F-89J "Scorpions" and the Convair F-102A "Delta Daggers." Another of their functions was furnishing navigational aid to friendly aircraft operating in the area. This was doubly important considering the severe arctic climate, and their detection of "May-Day" calls contributed to quick, efficient search and rescue efforts.

Forming the valley where the 64th Air Division (Defence) headquarters lay were Signal Hill and the White Hills with the tropospheric scatter antennas which provided the communication within the northeast area. Pepperrell Air Force Base, and the whole Island of Newfoundland, were steeped in both communication and aviation history.

The Red Cliff AC&W operation was similar to the operation at Jerry's Nose, except that more functional responsibility was given the station because of its close proximity to headquarters at Pepperrell. By the time Red Cliff closed in 1961, there were 140 military and 106 civilian workers at Red Cliff.

The cost involved by Canada alone in building the various AC&W stations was compiled in 1966, and included labour and materials. The US costs are not available and are believed to be much higher than those indicated below:

Red Cliff, St. Anthony, Frobisher, Cartwright, and other radar complexes	$100,000,000
Goose Air Force Base	$175,000,000
Ernest Harmon Air Force Base	$200,000,000
US Naval Station - Argentia	$150,000,000
Pepperrell Air Force Base	$ 90,000,000

The number of Canadian civilians employed in the construction of the various stations and sites was over 4,700. Many were rehired by the United States Air Force to maintain the station after construction. It is estimated over 3,000 civilians were employed at all the radar stations in Newfoundland and Labrador. The payroll to Canadian civilian employees, 95% of which were Newfoundlanders, amounted to over $23 million during the period of construction.

St. Anthony, Frobisher, Cartwright, Sagalek, etc.	$2,000,000
Goose Air Force Base	$7,000,000
Ernest Harmon Air Force Base	$8,000,000
US Naval Station - Argentia	$6,000,000

When the Red Cliff AC&W station closed in 1961, the US Navy at Argentia reconstructed and equipped one large building as an unmanned Radio Communications Centre. This was an elaborate operation, however, its usefulness was short lived when more sophisticated radio equipment came into use with long-range capabilities. Apart from specialized radar and radio equipment, all the assets, including the buildings and their contents, were turned over to Crown Assets for disposal.

The number of US military personnel assigned to each radar site depended upon its mission and what was required to keep it operational. Anywhere from 50 to 100 US Air Force personnel were stationed at each radar site. Tour of duty at extremely remote areas was one year; at other stations it ranged up to two years. The turnover of personnel was a constant problem and required cross training of all to help fill temporary vacancies. Job classification ranged from Radar Specialist to firefighter, heavy equipment operator, maintenance man, cooks and helpers to make up a self-sufficient organization. Civilian labour was used to the fullest where practical.

By 1960 it was apparent that the Pine Tree Line could be phased out. The introduction of long-range equipment, and the greater capability of refuelling long-range aircraft in the air, all brought about the change.

Although millions of dollars were spent on labour and equipment to establish a great North American defence organization, today the equipment is redundant. NORAD still operates out of Colorado Springs, and is still a US-Canadian operated organization. No one knew more than the Newfoundland people how much its shores needed to be defended from an aerial missile attack. There were times in the 1950s when Newfoundlanders were all on pins and needles, especially when jets had to scramble to identify a Russian aircraft near its shores.

Aerial view of the Canadian Forces Station, Gander, showing the two radar towers of the AC&W Station there.

10

Miscellaneous Major US Military Disasters

Those of us today may not realize the tremendous cost of life that was paid by American soldiers, sailors, airmen and Marines in and around Newfoundland and Labrador, from 1941 to 1985. Although many of the mishaps described in this chapter took place several years ago, and are not known to many today, it was certainly a supreme sacrifice for all concerned. During the Second World War years we can understand the tragedies which occurred, especially when we consider the enormous military activity that took place here. Even after World War Two, the movement of troops and supplies throughout US bases in Newfoundland and Labrador to support the Korean War, the Vietnam War, the Cuban Crisis and the scramble for security from Russia during the Cold War period, one can realize that the additional number of fatalities was a price paid by many. This chapter records over 600 American military personnel who were killed and, apart from the first Lockheed Bomber accident in 1941, deals only with American activity. We must also consider the number of RCAF, RAF, Canadian Army and Newfoundland civilian tragedies that occurred here over the years. The total would, in all probability, be an astonishing one. While many of the American military personnel were returned to the United States for internment during and after the war, many Americans are still buried here at the original fatal sites or elsewhere, or have never been found. All of the American servicemen who died in various accidents between 1940 and 1945 were buried in temporary military cemeteries located at Gander, Fort Pepperrell, Argentia, Harmon Field and Goose Bay. During 1946-47, the US Government had them all returned to the United States for burial with full military honours. We owe a great deal to those who died here during the war years and afterward, who in turn gave their lives for a great cause. *Lest we Forget.*

~ 1941 ~

★ In February 1941, an American-made RAF Lockheed Hudson Bomber III, number T9449, en route to England after take-off from Gander, and part of the Atlantic Ferry Organization, developed engine problems and crashed on a frozen lake near Musgrave Harbour, Newfoundland. Aboard the aircraft was Sir Frederick Banting, co-discoverer of insulin, who was killed, along with two others. Sir Frederick was en route to England on a secret mission regarding medical research. He was not killed in the crash, but received serious head wounds and died a few days later. The American pilot survived. Due to the isolated area of the crash site, it was three days before rescue parties arrived. Evidence of medical care given to some of the crew before they died substantiated the fact that Sir Frederick was alive after the crash.

Scene of the air crash in which Sir Frederick Banting was killed, Musgrave Harbour, Newfoundland, February 1941. A hospital in Gander, opened in 1941, was named the Banting Memorial Hospital in memory of Sir Frederick. (C-97C4)

~ 1942 ~

★ On 18 February 1942, two US Naval ships went aground at Lawn Point and Chamber Cove, near St. Lawrence on the Burin Peninsula. It was the worst marine disaster to occur along the Eastern Seaboard of Canada. Both ships were headed for the US Naval Station at Argentia when they ran into a fierce winter storm. The USS *Truxton* met her fate when she crashed ashore on a reef at Chamber Cove. The USS *Pollux* went aground at Lawn Point, about four miles from where the *Truxton* met her doom. The USS *Truxton* was a four master destroyer assigned to escort duty for the USS *Pollux*. It was similar in design to the fifty

The USS Pollux *was a supply ship carrying general cargo for delivery to Argentia at the time of her grounding. She had a total crew of 203 men aboard; only 140 survived.*

destroyers given to Britain in exchange for leases on American bases in the West Indies. Another US destroyer, the USS *Wilkes*, also went aground at the same time, but freed herself and continued on to Argentia in a damaged condition. Out of a total crew of 386 on both ships, 203 were killed and 183 rescued. The people of St. Lawrence and Lawn are credited with the rescue of these American sailors.

★ During the winter of that same year, at the height of construction of the US Army Signal Corps Long Line (Repeater) Stations across the Island of Newfoundland, tragedy occurred on Table Mountain, near St. Andrew's on the west coast. Because of the mountainous terrain and difficulties experienced in receiving and transmitting signals, the US soldiers assigned to this task had to select a site 2,000 feet high on top of a mountain to install special communications equipment. After three weeks all the materials had been hauled up the mountainside by manpower, and by sled using Siberian rescue dogs. Two men were then detailed to go back to the campsite to get some additional small parts, while the remainder started to set up the installation. On their return, and about a mile from the installation site, they were suddenly enveloped in a swirling snowstorm that quickly covered the trail. Afraid they would walk off the steep ridge of a cliff if they kept going, PFc Robert Malloy of Whitestone, NY and T-5 James Jasperse of Grand Rapids, Michigan huddled together behind a rock to wait out the storm. At dawn, search parties were sent out to find them. They found Malloy working his way around the mountain's crest, suffering from shock and exposure. He led them to the rock where he had taken shelter, only to find T-5 James Jasperse frozen to death.

- On 27 May 1942, the US Navy Destroyer USS *Prairie* caught fire at the Argentia pier and burned for over five hours. Two US sailors were killed.

- In December of that year another fire claimed 99 lives. The Knights of Columbus Hostel in St. John's was a favourite place for many, especially the military, for dancing and parties. On 19 December Uncle Tim's Barn Dance orchestra was performing and Biddy O'Toole was one of the singers when the fire broke out. The victims included sixteen RCAF, twenty-two Newfoundland Militia, seventeen RCNVR and Royal Navy, seven Canadian Army, nineteen civilians, five Merchant Navy, three US Army personnel and ten missing. The names of the US Army personnel from Fort Pepperrell were PFc Edward B. Ford, T-5 Henry L. Kennedy and Pvt. Frank M. Yirga. The cause of the fire was later to be determined as an act of sabotage.

- A US Army Air Force crew, flying from Greenland to Goose Bay, Labrador, crashed their B-26 bomber aircraft at Sagalek, Labrador on 10 December 1942. There was no military installation in that vicinity during that time. A blinding snowstorm and a northern "white-out" caused the crash. All six men aboard survived the impact with the ground, however, they were completely lost, with no radio and no knowledge of where they were. After spending two months in the wild, with little food and using a part of the plane as shelter, all six members perished. The irony of the event was that, unknown to them, they were only a few miles from the Eskimo village of Hebron. An Eskimo hunter found them around the last of February 1942. One of the crew kept a daily diary, recording all events of each day after the crash, to the day he died.

~ 1943 ~

- A major fire occurred at Gander in the late spring of 1943 when one of the Ferry Command hangars caught fire. Total destruction included an American B-24 Liberator bomber, and several other RCAF and RAF aircraft. Bombs aboard the planes exploded. There were no fatalities.

- On 4 August 1943, a USAAF B-17 Flying Fortress bomber aircraft crashed after take-off from Gander. Ten crew members were killed. Another B-17 crashed at Gander on 19 December; another ten crew members perished.

~ 1944 ~

- A B-24-H-20 USAAF aircraft, number 42-95146, crashed after take-off from Goose Bay, Labrador on 5 January 1944. The

The USS Pollux *as she lay wrecked on Lawn Point, 19 February 1942.*

The USS Truxton, *a four master destroyer. Out of a crew of 233, 137 lost their lives in the sinking of this ship.*

The remains of forty-eight victims of both the Pollux *and* Truxton *were buried at Argentia, and ninety were buried in the cemetery at St. Lawrence. After the war, their bodies were exhumed and returned to the United States.*

aircraft, en route to Europe, crashed in the Mealy Mountains. All ten crew members perished.

- On 8 February 1944 a B-17 Flying Fortress bomber crashed upon take-off at Goose Bay, and came down one-half mile from the airfield, in the middle of the large above ground gasoline storage tanks. It missed a one million gallon storage tank by 300 feet. All but two of the ten-man crew were killed, as the aircraft exploded and burned upon impact.
- On 19 April 1944, another USAAF B-17, en route to Europe, crashed after take-off from Goose Air Base, into Lake Melville; all ten members of the crew perished.
- A B-24L USAAF Liberator bomber crashed after take-off from Gander on 12 December 1944. The ten crew members aboard were killed.
- A fatal motor accident occurred on 23 May 1944 at Elliston, Bonavista Bay, when an American soldier fell from an Army truck. The soldier, a member of the 685th Air Warning Squadron, was engaged in the installation of a USAAF Radar Station at Cape Bonavista.
- A C-54 (DC-4) American Airlines aircraft, under contract to the USAAF, crashed during a landing approach to Harmon Field, Stephenville, on 11 December 1944. Nine crew members out of a total of eighteen died, including the pilot, Capt. E.C. Walkins.

~ 1945 ~

- A USAAF B-24-M-20 crashed north of Gander on 14 February 1945, while attempting a landing during inclement weather. The aircraft, which was returning from Europe en route to the United States, was not located until some time afterwards. All ten members of the crew were killed.

~ 1946 ~

- In February 1946 a fire at the RCAF Station at Torbay Airport destroyed Hangar #1 and several aircraft were destroyed, including a US Army Air Force plane. Bombs and other ammunition exploded.
- An American Overseas Airlines plane, en route to Germany through Harmon Field, crashed that October, about ten miles east of Stephenville. The New York to Berlin airliner rammed into the fog-ridden side of Hare Mountain. All thirty-nine persons on board were killed, including thirteen women and six children, all on their way to reunions with their American soldier husbands and fathers. The aircraft hit Hare Mountain about 100 feet from the top, exploded upon impact and burned. The site

On 11 July 1943 a large forest fire entered the town of Trinity and destroyed the Fisherman's Lodge and SUF Hall, a large store and several houses. The US Army at Gander sent several trucks to the area with men and equipment to help fight the fire. They also brought emergency food, blankets and other supplies.

This crash occurred at Nut Cove on 15 March 1953. A tail portion of the B-36 bomber aircraft and other debris in the foreground.

Fire at Pepperrell AFB, 20 March 1953, destroyed the new officers' bachelor quarters.

Hangar 1 fire at Torbay, taken the day after the incident. There were no fatalities.

has been known as "Crash Hill" since the date of the incident. All the victims were buried on the site and crosses erected over each grave. A three-foot-high headstone, located in the centre of the grave site, bears the names of all the passengers of the ill-fated flight. At that time, the crash was history's worst air disaster. In 1989, the RCAF 103 Rescue Unit stationed at Gander took on a project to make new grave marker crosses for all the victims, and install them on the Crash Hill location.

~ 1947 ~

★ On 10 December 1947 a C-54 transport aircraft with thirty-two USAF personnel aboard crashed and burned nine miles north of Goose Bay, Labrador. All of the passengers were en route to Westover AFB in Massachusetts to purchase Christmas gifts for the Post Exchange and NCO Club at Pepperrell and Goose. Twenty-three were killed and nine survived. Aboard the aircraft was M/Sgt. John Morris of Pepperrell AFB, who was killed. The first NCO Club at Pepperrell was named "Morris Hall" in memory of this outstanding non-commissioned officer. The structure, number T-851, still stands today and is one of only five remaining temporary buildings built at Pepperrell in 1942-43. The building is currently occupied by the Government of Newfoundland Crown Lands and Forestry.

★ Later that month, on Christmas Eve, a USAAF B-17-G-95DL crashed in Labrador. The aircraft was on a Christmas goodwill trip to outlying areas in Labrador, dropping Christmas gifts by low level flying. All the crew survived, however it was some time afterward before they were rescued.

~ 1951 ~

★ A USAF T-33 jet crash landed and burned off Runway 26 at Torbay Airport. No fatalities were reported.

~ 1953 ~

★ A USAF SB-29-70BW Search and Rescue aircraft out of Ernest Harmon Air Force Base crashed on 15 March 1953 in the Bay of St. George's. Only debris and an oil slick was found. There were no survivors of the ten-man crew. Three days later, a major crash occurred at Nut Cove, near Burgoines Cove (Clarenville) when a modern B-36H-25 Super jet bomber crashed on top of a mountainous hill. The aircraft was en route to Rapid City, South Dakota from Lagens Air Force Base in the Azores. There was a low ceiling and heavy fog conditions at the time. The aircraft was fully armed, however it did not carry any bombs. There were twenty-three aboard, including Brigadier General Richard Ellsworth, the Commanding Officer of Rapid City, SD Air Force

Base. All the personnel aboard were killed. The aircraft burned upon impact and parts of the bomber were scattered over a square mile area. The bodies were removed by USAF and RCAF rescue teams and some local residents. Parts of the aircraft are still in the position they were on that fateful day, including the ten engines of the aircraft.

- On 5 June a USAF F-94 jet type aircraft with two aboard developed a "flame-out" at 8,000 feet. The aircraft stalled, hit the ground and exploded just outside of Kippins, near Stephenville. Both pilots were killed. Just outside of Ernest Harmon Air Force Base, a C-54 transport plane crash landed and burned in March of 1953. All twenty-three aboard were killed.

- A US Navy C-121-R7BM Super Constellation aircraft out of Argentia crashed approximately eighty miles southwest of Harmon Air Force Base on 17 January 1955. Only some debris of the aircraft was sighted at sea. None of the thirteen naval members aboard survived. A major fire at Goose Air Base on 10 April 1955 caused destruction of a large hangar and several USAF aircraft.

~ 1956 ~

- A USAF T-33 jet aircraft, piloted by Col. Carl Payne of Pepperrell AFB, crashed into a house at Torbay on 9 January 1956. The aircraft developed a "flame-out" and Col. Payne was killed.

~ 1958 ~

- In October 1958, a US Navy E121 Superconnie patrol aircraft crashed at Argentia during the aircraft's second landing attempt. It fell 1,000 feet short of the runway in thirty feet of water. The eleven man crew was killed. The aircraft was involved in submarine and northern intrusion patrol. This was the second aircraft of this type to crash in a seven month period.

~ 1959 ~

- Yet another E121 Superconnie patrol aircraft crashed at night at Argentia on 4 February 1959. The aircraft was attached to the AC&W Squadron Number 13. It was returning to the base, thirty-five minutes after take-off, with engine trouble. The aircraft was making a ground control approach (GCA) in foggy conditions when the landing gear collapsed upon contact with the runway. Out of the twenty-four man crew, one was killed and all others injured.

- On 23 April 1959 an F-102 jet fighter of the 59th Fighter Interceptor Squadron at Goose Bay crashed in a bog near Springdale. Due to high winds at high altitude, the pilot, Capt.

Aubrey Seal of Jackson, Mississippi, bailed out and landed near Fortune Harbour. Parts of the pilot's ejection seat were found in 1960 during a woods' survey.

★ A USAF B-47 bomber crashed at night at Goose Bay, Labrador on 2 October 1959. The aircraft was from the 310 Bomb Wing at Schilling Air Force Base in Kansas. Two crew members were killed and two seriously injured.

~ 1960 ~

★ In 1960, a crash occurred in Port au Port Bay, involving a F-102 Interceptor jet fighter out of the 59th FIS detachment at Goose Air Base. The pilot, Lt. Richard Lyttle of Indianapolis, Indiana, was picked up by a H-19B helicopter. The aircraft experienced a "flame-out."

★ A USAF C-124 Globemaster aircraft crashed one mile from Harmon Air Force Base after take-off. The aircraft was from Dover Air Force Base in Delaware and was en route to the Azores with medical supplies and mail. The aircraft caught fire upon impact and burned. Seven crew members perished.

★ Fire at the elaborate USAF Officers' Club at Goose Air Force Base caused total destruction on 14 July 1960.

~ 1961 ~

★ On 26 January 1961 a C-118 USAF Military Air Transport Service aircraft crashed off the Newfoundland coast, approximately thirty miles south of Argentia. All ten members of the crew and thirteen military passengers were lost. Only an oil slick was located. The aircraft was en route from Europe to Argentia for a stopover.

~ 1963 ~

★ A U.S. Navy C-121 Super Constellation aircraft crashed at Gander on 30 June 1963 while attempting a landing. The aircraft hit the threshold of Runway 22 only 100 yards from the tarmac. Only one of seven crew members was injured.

~ 1964~

★ A USAF C-133B cargo aircraft crashed after take-off at Goose Air Base on 7 November 1964, near the large gasoline storage tanks. All seven members of the crew were killed. The aircraft was en route to Sondrestrom Air Base in Greenland with supplies.

★ A US Naval P3A Orion type aircraft crashed in approximately one hundred and ten feet of water off the end of the active

Officers' Club fire at Goose Bay, July 1960.

runway at Argentia on 15 November 1964. None of the ten US Navy crew members survived.

~ 1985 ~

★ The most recent military disaster occurred on the morning of 12 December when an Arrow Airlines chartered US military aircraft departed Gander International Airport with 248 US soldiers and a civilian crew of eight aboard. The aircraft crashed and burned after take-off on the Trans Canada Highway side of Gander Lake. All aboard were killed. The soldiers were en route to Fort Campbell, Kentucky on Christmas leave. They had departed the occupied West Bank of the Sinai on December 11, where they were serving with the 101st Airborne Division on duty with the UN Contingency Forces. It was the worst air disaster to occur in Newfoundland and on Canadian soil.

Artist's drawing of proposed Samaritan Place, Gander, which will be constructed as a living memorial to those who died in the 1985 Arrow Arlines Crash.

11

Memorable Events with the Americans

It has been said many times that "no one knows the overall benefits the Americans brought to us in Newfoundland and Labrador." It stands to reason that many exchanges of friendship must have occurred in five decades of association. Many do not know of the many things the Americans did which were above their primary mission of protecting the Island against enemy attack, and supporting the war effort. To itemize each and every benefit and goodwill accomplishment given by them would require a great deal of space to record. Therefore, for the information of those who are and were too young to realize the changes in our way of life that occurred, particularly from 1940 to 1960, the following paragraphs are a general overview.

Billions of dollars were spent to construct and operate the American military installations in this province which contributed to the employment of over 20,000 Newfoundlanders and Labradorians and to the construction and maintenance of new and existing roads. Friendships developed between the Americans and the Newfoundlanders, especially through sports and recreation events. Because of the USO, many Newfoundlanders got the opportunity to see, and be entertained by, famous movie stars and performers. The American military made significant contributions to Newfoundland charities and was closely affiliated with various churches, choirs and religious organizations. There were over 25,000 marriages performed in these churches between American military personnel and Newfoundland women. Hundreds of orphaned babies were adopted by American personnel and the military also organized and conducted various youth programs. Besides the many personal relationships, the Americans fulfilled their prime directive and kept Newfoundland and Labrador secure during the war years. In the war era Newfoundland seemed like the newest

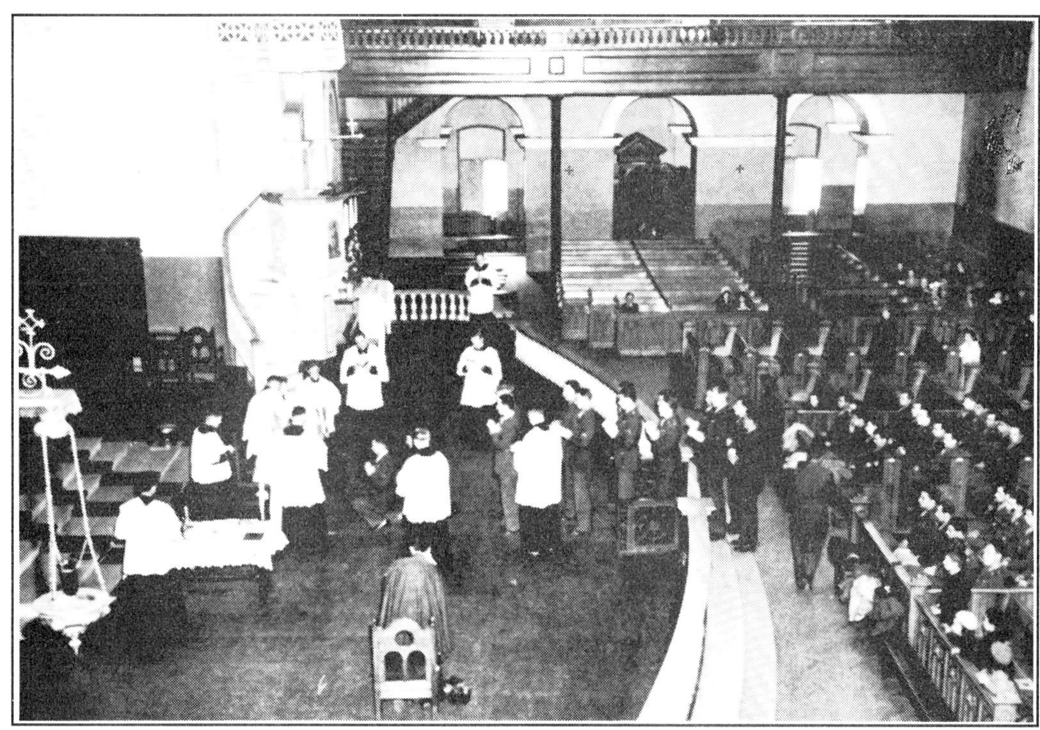

Sanctuary view of the confirmation of US soldiers in the Roman Catholic Basilica in St. John's, 1943.

One of the many weddings in the base chapel at Fort Pepperrell. This is the 1951 wedding of Capt. Joseph R. Smith and Ann Tizzard of St. John's. Maid of Honour is Miss Philomena Cleary and Best Man is Leo Shea. The name of the chaplain is unknown.

Archbishop Roche administering the Sacrament. Fathers J.W. O'Mara and Ronald Murphy are in attendance.

state in the Union to many American servicemen. Yet, these Americans quickly learned the Newfoundland way of life and became used to such phrases as "Come 'ere, me son!" "Yes, b'y," "dare say," "I 'llows," "Hi ya, me darlin'," "How ya gettin' on, b'y," and "Stay where yer to, 'till I comes where yer at."

American soldiers, sailors, airmen and Marines who were stationed here, and their numerous Newfoundland friends, have collected many memories of happy and historic events. While it would be difficult to include all of the photographs and other paraphernalia collected, the events identified in this chapter will, no doubt, bring back many happy memories.

After the war, in 1947, the US Navy at Argentia decided to sell some of its small boats. Mr. Walter H. Davis of St. John's, on behalf of the TB Association of Newfoundland, proceeded to Washington, DC and acquired the US Naval Aircraft Rescue Boat, the PT-107. (This was a sister boat to the PT-105 commanded by the late US President, John F. Kennedy, during the war.) The US Government, through the US Navy, sold the boat to the Newfoundland government for $1.00. It was intended to use the boat in the Newfoundland Tuberculosis Campaign throughout the Island. The PT-107 was renovated and refitted in St. John's and was named the MV *Christmas Seal*. Newfoundlanders will recall the great work done by this small vessel since 1947 in the fight against tuberculosis. The vessel called into hundreds of ports along the coast of Newfoundland and Labrador, and the doctors and nurses aboard X-rayed over 500,000 people. X-rays were repeated for many suspected with the disease. It would take the MV *Christmas Seal* three years to complete its round of calls to Newfoundland and Labrador ports. Captain Peter Troke, Captain Carl Barbour, Captain Harding, Captain Laite and Captain Fraize Windsor were some of the captains from 1947 to 1969, when the MV *Christmas Seal* was retired from service, after serving all this time with the Department of Health. The vessel was purchased by Mr. Clyde Mullett of Lewisporte in 1979. While on a voyage to Sydney, Nova Scotia, fire broke out aboard the boat and it sank.

Jack Brown and helper at the PX ice cream plant, opened in 1945 at Pepperrell.

Starting in 1941, and continuing until 1960, the American Red Cross maintained a headquarters at Fort Pepperrell, and a full office at each major US military base. The American civilian men and women who manned these posts served all the military and civilian employee personnel in many ways. Some of their duties included: arranging for leave and transportation to the States for soldiers or sailors, where a serious illness or a death occurred in a family; assisting families in the States in locating their son, husband or father in the service; coordinating with the American Consulate in St. John's and their American Red Cross headquarters in Washington, DC, where a problem existed with a soldier's or sailor's family;

Top left: The PT-107 *coming into Argentia after a short voyage in winter seas. The* PT-107 *was assigned to Argentia Naval Station in 1942, and was responsible for the rescue of many aircraft crew members who crashed in the vicinity of Argentia.*

Centre left: MV Christmas Seal *on her arrival in St. John's, 17 November 1952, having visited every navigable port in Newfoundland and Labrador. The MV* Christmas Seal *was originally the* PT-107 *Rescue Boat at Argentia.*

Bottom left: Private party and dance in the dayroom of Building 808, 28 December 1942, to celebrate the award of a Good Conduct Medal to Pvt. Russell Englund of the US Infantry. Pvt. Englund (dancing with wife at right) arrived on the Alexander *and was stationed at Pepperrell for over twelve years. He retired in 1964 at the rank of 1st Sgt. (Master Sergeant) and has lived in St. John's ever since. He was married to Olive Brake of St. John's.*

Bottom right: Dance held at the opening of the large Base laundry building at Fort Pepperrell in 1943. This structure today is used as the Central Laundry for all hospitals in St. John's and surrounding areas.

providing emergency services to Newfoundland civilian employees serving in isolated locations on American bases, and many more personnel services. Another of their many duties was to ensure the health and welfare of the American military. Many of the Red Cross ladies organized dances, parties, picnics and other entertainment from 1941 to 1945, on the base where they were assigned. This work, to boost the morale of the military, did not stop or even diminish when the United Service Organization (USO) established itself in St. John's and Corner Brook. The American Red Cross headquarters in Newfoundland closed its office in June 1961.

Many US soldiers of the 8th Weather Squadron and the 136th AACS Squadron served at Eureka, Nord and Alert in the Arctic, as well as at Fort Chimo and Mingan, Quebec, Padloping Island on the Cumberland Peninsula, Indian House Lake, Quebec and Cape Harrison, Labrador. All of these stations were very remote and could only be serviced from Newfoundland and Labrador by aircraft.

During the war years many famous actors, singers and other dignitaries served with the US Forces in Newfoundland. Actors Victor Mature, Hal Holbrook and singer Steve Lawrence served at Argentia. The ABC News anchorman on Detroit's Channel 7, Bill Bonds, served at Pepperrell. Johnny Williams, who composed the score for "Star Wars," and now conducts the Boston Pops Orchestra, served at Pepperrell in the 596th USAF Band Squadron. The chef at the famous Waldorf Astoria and Westchester Country Club in New

Personnel of Starlighters Band are:—T/Sgt. Jim Wallace, leader; A/1C Lois "Torchy" Barcroft, vocalist; Sax Section:— A/2C Gerry LaBombard, baritone; A/3C Frank Hittner, tenor; A/1C Gordie Heckel, alto; S/Sgt. Randie Rogers, alto; A/3C Dave Sand, tenor; Trombones:—A/1C Bobby Allen; A/2C Al Weiner, S/Sgt. Roger Dengler; Trumpets:—S/Sgt. Bill Rowe, S/Sgt. Bob Cummings, A/2C Tom Miraglio, A/1C Dale Nelson; Rhythm:—A/2C Johnny Williams, piano; A/1C Cliff William, bass; A/1C Jack Kohr, drums.

The USAF "Starlighters" Band performing at the Old Colony Club, St. John's, 1953. The dance was a held to raise funds for the St. John's Memorial Stadium. Note A/2C Johnny Williams on the piano (far left), renowned for his work with the Boston Pops Orchestra.

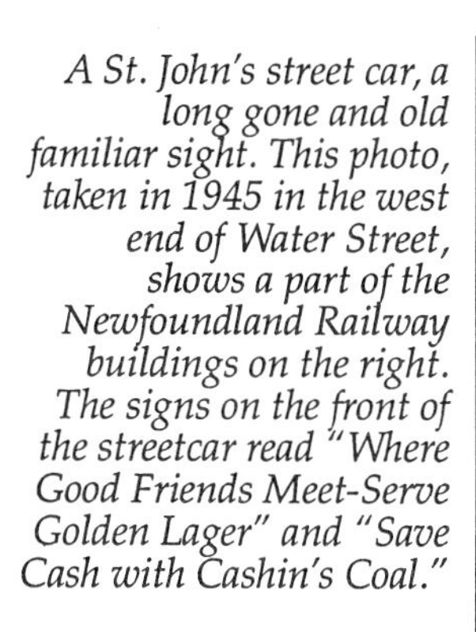

A St. John's street car, a long gone and old familiar sight. This photo, taken in 1945 in the west end of Water Street, shows a part of the Newfoundland Railway buildings on the right. The signs on the front of the streetcar read "Where Good Friends Meet-Serve Golden Lager" and "Save Cash with Cashin's Coal."

The First Post Exchange (PX) at Harmon Field in 1942 at Camp Morris showing Owen Dunphy and Paul Kavanagh of Stephenville.

"B" and "D" Battery canteen on Signal Hill, 1944.

A group of soldiers from Battery "B" in front of the Colonial Building, near Bannerman Park, 1943.

York, served as a Mess Sergeant in the US Army at Gander. Clarence Englebrecht, well-known local TV and radio personality Bob Lewis, served as a captain at Pepperrell. Ralph Walker, local pianist/entertainer, served with the 596th USAF Band at Ernest Harmon Air Force Base at Stephenville and St. John's.

US military bases in Newfoundland and Labrador were never called by their names or locations during the war. Each base was identified with a code name:

APO 862 - Fort Pepperrell	Crystal I - Fort Chimo, Que.
APO 863 - Fort McAndrew	Crystal II - Frobisher, NWT
APO 677 - Goose Bay	Crystal III - Padloping Island
APO 864 - Harmon Field	BW 1 - Narsarssuak, Greenland
APO 865 - Gander	BW 8 - Sondrestrom, Greenland
FPO 09519 - Argentia Naval Station	

APO was the abbreviation for Army Post Office, and FPO was the Naval designation for Fleet Post Office. All mail for APO and FPO located overseas went to New York for sorting and rerouting. After V-J day in August 1945, the names and locations of US bases could be used.

It has been said that when the US Naval Air Station at Argentia and the Atlantic Ferry Command operations at Gander, Goose and Harmon were at their peak of operation in 1943-44, Field Marshall Goring of the German *Luftwaffe* said to Hitler: "The Island of Newfoundland is like a giant aircraft carrier, and must somehow be destroyed."

Before the Argentia-Holyrood road was completed between Argentia-Colinet-Markland, many pieces of heavy equipment and supplies from St. John's were transported by freight train to Whitbourne. They were unloaded, started and driven down through Markland to Argentia. To this day, buried in deep bog, is a large crane that evidently went off the road and sank in the bog. A few years ago a resident of Whitbourne found the location by driving a metal rod into the bog until it hit a part of the crane.

When United States President John F. Kennedy was assassinated on 22 November 1963, Newfoundland and Labrador was shocked and in a state of mourning. People were seen crying on the streets; some flags flew at half-mast on businesses and at private homes. The effect on the people of Newfoundland was expressed in many ways. Strangers met American soldiers on the street and shook their hands in an expression of sorrow. Every person who possibly could was listening to their TV or radio for word of hope. Many local engagements for that night were cancelled. The closeness between the Newfoundland and American people was very clearly demonstrated on that fatal day.

The new blood donor room of the Canadian Red Cross at their headquarters on Wicklow Street, St. John's. The room is called "Narsarssuak Room."

Newfoundland's first CBC-TV station went into operation at Harmon Air Force Base on 1 March 1957. The station was jointly operated by the Canadian Broadcasting Corporation and the US Government. The station supervisor was Mr. Arthur Barrett of St. John's who was the CBC representative, 1st. Lt. Eugene Hunguford, officer in charge and M/Sgt. Adolph Bryson, the station non-com, with seven other technicians. The station operated on Channel 8, with a power of 200 watts and a fifteen-mile radius coverage. The station was housed in the base Rod and Gun Club, with a control room, telecine and film editing room, studio, film projection room and offices. The station was called CFSN-TV and served Harmon and surrounding area for many years.

Each major base had its own military newspaper or magazine. Some of these have been identified as:

Argentia Flyer - Argentia US Navy, 1980-1989
Harmoneer - Harmon Field, 1942-1945
Harmon Flash - Ernest Harmon Air Force Base, 1950-1958
Proppagander - Gander (Quarterly), 1942-1945
The Prop - Gander (Daily), 1942-1945
Command News - Pepperrell Air Force Base, 1942-1945
Northeast Guardian - Hq. NEAC, Pepperrell, 1953-1957

News 'N Blue - Pepperrell Air Force Base, 1950-1957
Torbay Times - Torbay Airport, 1953-1958
Vanguard - 64th Air Division, Pepperrell Air Force Base, 1958-1961
Beacon - Goose Air Base, 1955-1958
GAB - Goose Air Base, 1960-1964

Every edition carried items on the civilians who worked on the base.

When the Narsarssuak American Air Force Base in Greenland closed in 1957, civilian employees who operated the "Newfoundland Civilian Club" donated $77,000 to the Canadian Red Cross in Newfoundland. Mr. L.M. Parsons, Commissioner for the Red Cross in Newfoundland at the time, stated that the money was used to establish the first Blood program and Blood Bank in Newfoundland. Prior to that date, Newfoundland was the only province in Canada that did not have a Blood program or Blood Bank. The new Blood Donor program and Blood Bank opened in 1958 at the old Red Cross building (now Harvey's Travel) in Devon Row on Duckworth Street east. The building was named "Narsarssuak House." The building maintained this name until the Canadian Red Cross moved into its new headquarters on Wicklow Street, St. John's, in October 1979. Today, the Blood Donor Room and Blood Bank is called "Narsarssuak Room." Miss Helen Dunn, the director of the Blood Donor Recruitment program, served with the American Red Cross headquarters for three years at Pepperrell Air Force Base. Prior to 1949 the British Red Cross was in Newfoundland.

From the day Goose Air Base was opened in 1941 and up to 1945, only four women worked and lived on the base. They included two American Red Cross ladies, one American nurse and the RCAF Commanding Officer's secretary. There were over 5,000 military and

During the war years each civilian who worked on a US military installation was required to have his or her identification badge at all times. These two photos show Walter H. Davis' badge when he worked at Fort McAndrew in 1942, and Leo Kerwin's when he was employed at Fort Pepperrell in 1949. These badges were later replaced by plastic laminated passes.

Mr. Leo Shea, from HQ, NEAC, Pepperrell, enjoys a good steak at the Raven's Roost Airmen's Club in Narsarssuak, Greenland, in 1955. He is sitting with Major George Stanton and Club manager T/Sgt. James Crawley of Texas.

A group of Newfoundlanders at Narsarssuak Air Force Base, Greenland ,who voted for the contribution to the Canadian Red Cross. They posed for this photo in the Narsarssuak Newfoundland Civilian Club in 1950.

The Canadian Red Cross building in St. John's was named the Narsarssuak House in 1957. The first Newfoundland blood bank and blood donor program began in 1958. Here, local advertising promotes the blood donor program. From the left, are: Al Eaton, Norman Duffett and Bill Squires.

over 700 civilians at Goose Bay during peak activity periods. On the American side, over 3,000 American troops were assigned in 1945. Needless to day, morale was boosted to 1,000 percent when, during the summer of 1945, forty USAAF WACS arrived for duty at Goose Bay. In the early fall, thirty-two civilian ladies recruited by the Department of Labour, Moncton, New Brunswick also arrived for miscellaneous duties in the Post Exchange, Officers' and NCO Club and Coffee Shop. The WACS served at Goose for one year. In 1947 civilian female workers were hired for various duties on this base. On 30 June 1949 forty-nine USAF WACS arrived at the Army Dock, St. John's, aboard the UST *Sergeant Jonah E. Kelly*. They remained at Pepperrell for over a year. US military servicewomen also served at Argentia and at Harmon Field.

Many American soldiers, sailors and airmen remember many local orchestras and bands besides Chris Andrews, during the war years and after. Some were Uncle Tim's Barn Dance, Mickey Duggan, Bob McLeod, Leo Michaels, Barry Hope, Irene B. Mellon and Miles Frelic. In the Argentia area they had the Commanders, Gordon Noseworthy's Band and the Smith Brothers. Local singers of these days were Biddy O'Toole, Nellie Ludlow, Shirley Small, Margie Clark, Jesse Earles and Isabelle Rickitts. They still talk about their favourite nightclubs or dance halls in St. John's, such as the Caribou

The combination band of T/Sgt. Loy (Red) Foster at the USO building, 1946. Note members from the US Army, Navy and local civilians. Red Foster also conducted the US Army Air Force Band.

Hut, Old Colony Club, Crystal Palace, Guard Club Room, K of C Hostel, USO building, Old Mill, Piccadilly Club, Commodore Club, Gerry Byrnes, Club 21, SUF Hall, Newfoundland Hotel Ballroom, Bella Vista and Morris Hall (NCO Club) at Pepperrell. The Red Rose Club, Polar Club, Dune Lodge, Wheelers, and Frenches were popular at Stephenville, while the Glynmill Inn, the K of C Hall and the USO Building were the hot spots at Corner Brook. Grand Falls had the K of C Hall and there was the Brass Rail in Clarenville. In the Argentia area there was the Pink Lady in Jerseyside, Latin Quarter and Hillview in Placentia. The "Skymasters Swing and Sway Band" at Harmon Field were members of the USAAF Band. They entertained thousands at Harmon and at the USO in Corner Brook.

Remember the great American songs of the early and late 1940s which were played over local radio stations and especially requested at local dances? To each person, each song will recall beautiful memories.

'Til the End of Time
As Time Goes By
Give Me Five Minutes More
Don't Fence Me In
If I Loved You
This is the Army, Mr. Jones
I'm Beginning to See the Light
It's Been a Long Long Time
Candy
Chickery Chick
Atchison, Topeka and the Santa Fe
Begin the Beguine
Doin' What Comes Naturally
Looking Over A Four Leaf Clover
Dark Town Strutters Ball
Baby It's Cold Outside
See You In My Dreams
Don't Sit Under the Apple Tree
Slow Boat to China
Rosie the Riveter
A Wonderful Guy
Stardust
Serenade in Blue
You're Breaking My Heart
Forever and Ever
Hey! Ba-Ba-Re-Bop
Over the Rainbow
Now is the Hour
Elmer's Tune
Temptation
Red Roses for a Blue Lady
You Call Everybody Darlin'
Dancing in the Dark
To Each His Own
Dig You Later
Cement Mixer
Laughing on the Outside
That's My Desire
Anniversary Song
Heartaches
Oh! What it Seems to Be
South America - Take It Away
For Sentimental Reasons
Peg O' My Heart
Manana - Soon Enough for Me
Now is the Hour
Deep Purple
Room Full of Roses
I Belong To You
Some Enchanted Evening
Far Away Places
Bell Bottom Trousers
It's A Good Day
Tea For Two
Mule Train
Bali-Ha'i
Little White Lies
Making Believe
My Happiness
Blue Moon
I Can't Begin to Tell You
Caladonia
Long Ago and Far Away
Laura
Blues in the Night
Stormy Weather

Miss You Since You Went Away
That Old Black Magic
I'll Walk Alone
White Christmas
Rock-a-Bye Baby - Dixie Melody
Wash that Man Right Out of My Hair
My Dreams are Getting Better all the Time
Smoke - Smoke - Smoke That Cigarette
I've Got My Love to Keep Me Warm
Our Love is Here to Stay
Embraceable You
The Nearness of You
Easter Parade
Saturday Night

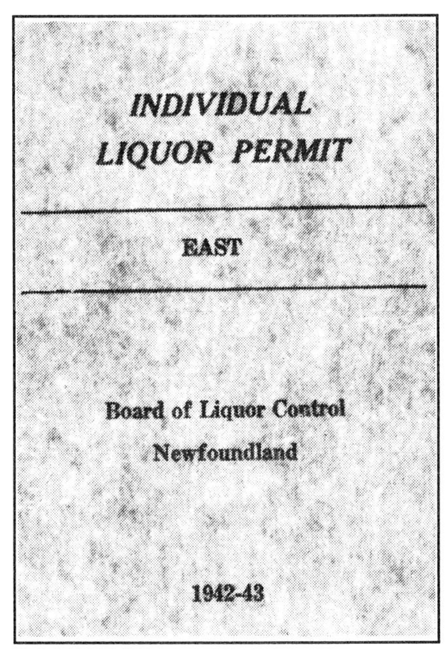

This liquor book, from 1942-43, belonged to Pvt. Peter E. Steffens of Bayville, New Jersey, who served at Fort Pepperrell during those years. His last purchase was for two bottles on 8 July 1943. Liquor books were first used in Newfoundland in 1941.

Former American servicemen will recall their favourite rendezvous areas, such as Bowring Park, around Quidi Vidi Lake, Camp Alexander area, Victoria Park, Water and Duckworth Streets, Signal Hill and Cabot Tower, Bannerman Park, the Southside and the Waterfront. Stephenville had the Beach area and Port au Port. Corner Brook had the Bowaters Park and Steady Brook. They also remember the cobblestones on Water Street, the street cars, Golden Arrow Coach buses, funerals with a horse and hearse, driving on the left-hand side of the road, the Newfoundland Railway station, the Furness Witty line at the wharf, the War Memorial, the Gerald S. Doyle news bulletin and CJON TV. Some of the downtown businesses for years carried slogans such as "Once a number, now an institution," "Up the line with Jackman and Green," "At the Sign of the Book" and "We save their soles and heel them."

The US military did not have the privilege of buying bottled liquor on the bases until 1946. Like everyone else, they had to purchase a liquor book from the Board of Liquor Control (Controllers) and each person was allowed only two bottles a week. Some of the liquors sold from 1940 to 1946 were: Demerara 1R and 1S rum; London Dock rum; Old Niagara wine; Haig Ale; Moose Ale; Indian Pale Ale; Golden Lager; Black Horse; and Guinness Stout. The old Jamaican rum, bottled in St. John's, was nicknamed "Newfoundland Screech" and is today sometimes referred to as "Newfie" Screech, a nickname the Americans coined in 1942.

Throughout Newfoundland and Labrador, over 25,000 Newfoundland girls married US servicemen. If we were to take one-half of these families and estimate that each family had at least two children, we are talking about 30,000 American children with Newfoundland mothers. Since 1940, many of these children have grown and are married with children of their own, thus thousands of American people can trace their ancestry to Newfoundland.

The American military at all US bases and sites throughout our province contributed thousands of dollars to our charitable organizations. The American Legion's records indicate many Christmas donations were made to the orphanages in St. John's. In 1950 both Mr. Walter N. Davis and Mr. F.M. O'Leary coordinated activities with the Commanding Officers of all bases, on the TB

Left: Leaving the chapel after the 1951 christening of Baby Bleau, daughter of Captain Bleau of Pepperrell. Left to right, Mr. and Mrs. Vince Wiltshire (godparents), Ronnie and Deanne Bleau, Mrs. Bleau with baby, and Margie Wiltshire.

Right: The start of the 1953 "Mile of Dimes." Shown left to right are: Stadium Council representatives T. Whitten; Ank Murphy; Jim Tucker; St. John's Mayor Harry Mews; F.M. O'Leary; Al Andrews; and Doug Oliphant. At the far right is Col. Roy D. Butler, Base Commander, with several commissioned and non-commissioned officers, members of thePepperrell Mile of Dimes committee.

activities with the Commanding Officers of all bases, on the TB Association drive, resulting in yearly substantial contributions from US military personnel. This program continued until the bases closed. Campaigns were conducted on all the US bases for the Canadian Red Cross, the March of Dimes, the Canadian Institute for the Blind, the Cancer Society, and many more. In 1955 the US military in the Northeast Air Command collected over $22,000 for the Red Cross. Operation Santa Claus was a highlight in which US helicopters transported Santa to anxiously awaiting children in the city of St. John's and on all the US bases. In 1953 a scheme to raise funds for the new St. John's Memorial Stadium was coordinated between the Stadium Committee and the Commanding Officer of Pepperrell AFB. A line (one mile long) was painted on the road from the Stadium down The Boulevard to the main gate at Pepperrell. It was called the "Mile of Dimes." The object was to collect enough dimes that could be laid end to end on the mile-long course. The American military and civilian employees at Pepperrell met this goal. In 1955 the Commanding Officer of Goose Air Base, Col. J.B. Knapp, raised a considerable sum of money for the Grenfell Mission in Labrador. He had a slot machine placed by each cash register on the base for one month and donated all proceeds.

the news. In 1956 the "Pepperrell Peppers" baseball team, out pitched, out hit, and out fielded all the armed forces teams in Iceland, Greenland and Newfoundland and Labrador, to win the Northern Overseas Baseball Tournament. All of the daily and quarterly newspapers printed on each US base featured articles on such sports as football, baseball, softball, soccer, tennis and basketball. Many competitive games were played between the Americans and Newfoundlanders.

The Office of the United States Consulate in St. John's was opened on 4 October 1852 by William Newman, who was the first US Consul. Mr. Newman was born in Newfoundland of English parents. He emigrated from England to New York, and subsequently became a naturalized American citizen. He served as US Consul in Newfoundland until his retirement in 1862. Altogether, since 1852, twenty-seven persons have been appointed as the US Consul or Consul General in Newfoundland. The following is a list of these consuls and the dates they served here:

William S.H. Newman	October 4, 1852 - May 27, 1862
Convers O. Leach	May 28, 1862 - July 12, 1865
Thomas N. Malloy	July 13, 1865 - February 1, 1895
Sam Ryan	February 2, 1895 - October 1, 1895
John T. Barron	November 14, 1895 - October 31, 1897
Michael J. Carter	September 1, 1897 - May 8, 1901
Henry F. Bradshaw	January 21, 1901 - November 12, 1901
George O. Cornelius	November 12, 1901 - September 19, 1907
James S. Benedict	September 20, 1907 -July 1, 1924
Avra M. Warren	July 2, 1924 - July 10, 1930
Edward A. Dow	July 11, 1930 - May 31, 1933
George K. Donald	April 7, 1933 - June 20, 1934
Harold B. Quarton	June 15, 1934 - June 9, 1941
George D. Hopper	June 19, 1941 - September 17, 1945
George K. Donald	November 26, 1945 - June 30, 1947
Wainwright Abbott	June 4, 1947 - July 28, 1949
Sidney A. Belavaky	May 5, 1949 - April 9, 1951
Horatio Moores	November 20, 1950 - February 5, 1951
William E. Flournay	September 23, 1954 - July 15, 1956
Kenneth A. Byrns	July 16, 1956 - June 15, 1958
William H. Christensen	June 30, 1958 - September 18, 1961
Homer W. Lanford	August 5, 1961 - December 31, 1966
Russell R. Riley	August 29, 1966 - July, 1968
Richard Strauss	July 1968 - July 1972
Theodore B. Dobbs	June 1972 - June 30, 1976

Two other persons appointed declined the position. During all times when the US Consulate did not have an appointed Consul, the Vice-Consul of the time assumed all duties.

The office of the US Consul looked after all the diplomatic affairs of the United States while Newfoundland was a separate colony and after Confederation with Canada. The office of the US Consul was in the Commercial Chambers building on Water Street from 1897 until 1952. On May 21, 1952 a contract was signed to begin construction of a government owned US Consulate General office building and residence on King's Bridge Road. The site had been leased on 21 June 1946 from the Diocesan Synod of Newfoundland for an initial cost of $20,000 and a yearly rental fee of $1.00 for ninety-nine years. The new office and residence were occupied in the summer of 1947. The office was closed on 30 June 1976 and relocated to Halifax, Nova Scotia. The building on King's Bridge Road was sold to the Diocesan Synod and the land returned. The structure is today occupied by them. Many Newfoundlanders will recall the attractive US Consulate Building on King's Bridge Road, next to the east gate of Government House. The front lawn had a beautiful garden setting, a half-circle driveway and the seal of the United States on the front of the building. The garden at the rear was a very spacious and beautiful one, to which many Newfoundlanders were invited for receptions.

On 9 July 1988 a reunion of all former civilian employees of Goose Bay Air Force Base was held at the Royal Canadian Legion, Branch 56 in St. John's. All attending were honoured when Lt. General James B. Knapp (Retired), Base Commanding Officer from 1953 to 1957, attended the reunion and banquet with his family. Over 500 attended the two-day affair. The first reunion was held in St. John's in 1983 with over 600 in attendance.

Over $8,000 was won in 1959 by over sixty-eight civilian employees at Pepperrell Air Force Base and Command Headquarters for sustained performance on the job. Thirty-seven civilians won cash awards for suggestions regarding improvement of working conditions on the base. Over 450 suggestions were submitted. The Meritorious Civilian Service Award for 1959 went to Mr. Chancy Greer, Commissary Officer. The highest civilian award which could be approved at a major command level went to Mrs. Avery Collins, school teacher at the base school, and to Mr. Ronald Tilley, Assistant Director of Budget for the 64th Air Division. Civilian awards were given every year at each Air Force Base in Newfoundland and Labrador.

US Army personnel were on special 24-hour guard duty from 1941 to 1945 at many strategic points in Newfoundland. One was the dam at Millertown Junction; another was the gasoline and oil storage on the Southside Hills and Windsor Lake area. The main road bridge at Colinet was mined with high explosive charges that could be detonated from the US Army campsite in Colinet.

From as early as 1939, when Newfoundland, as a British colony, became actively engaged in the Second World War, all radio sets occupied by Newfoundland and Labrador residents had to be licensed. The license fee for use of a radio was $2.00 per year. This rule did not apply to US or Canadian soldiers who resided on military bases. However, those who resided off base had to purchase a radio license.

Interior of the Post Exchange at Fort Pepperrell, 1948.

Canadian and American civilians of the 64th Air Division received cash awards for outstanding performance in 1959. Left to right, front row: Mrs. Muriel Tucker, Miss Ellen Walsh, Mrs. Elayne Judge, Mrs. Ruth Jackman, Mrs. Mary Jo Howard. Back row: James Barton, John Cardoulis, Mrs. Cathleen Fitzgerald, Miss Doris Harvey and John Weidman.

Civilian employees attached to the 4737th Air Base Group, Pepperrell Air Force Base, received cash awards for outstanding and sustained work performance, 30 December 1959. Left to Right, front row: Mrs. Dorothy Price, Miss Ann O'Brien, Mrs. Mary Evans, Miss Emma Elton, Miss Mary Hannaford, Miss Margaret Cadigan, Mrs. Johanna Kennedy. Back row: Alfred Smith, Leslie Cullomore, Herb Knight, Jacob Tucker, Freeman Stanley, Charles Henley, John Connors, Arthur Hackett, Gerald Healy and Gordon Tizzard.

On August 19, 1939, the President of the United States, Franklin D. Roosevelt, visited Corner Brook as a guest of Bowaters Newfoundland Pulp and Paper Company. Arriving on his flagship, the *Tuscaloosa,* he was greeted at the paper mill pier by several hundred people, many of whom were American citizens. The President spent some time salmon fishing at Steady Rapids, where his aide-de-camp, General Watson, landed an eighteen pound salmon.

The Town of Stephenville organized and celebrated a "Yank Come Back" campaign during the summer of 1986. A three-day program of activities from 2 August to 4 August highlighted the US military at Ernest Harmon Air Force Base. Hundreds of American ex-servicemen returned to take part in the celebrations. The USAF sent in several of their latest fighter and bomber aircraft for an air show, and for open display to the public. The American Legion conducted a "Yank Come Back to Newfoundland" campaign in 1988 as their part in the "Great '88 Soiree" program. Thousands of letters were received by the Legion in response to advertising in the *American Legion* magazine. Over 2,500 former US servicemen who were in the American Forces from 1940 to 1976 and served in either Newfoundland or Labrador returned to the province during the period of May to September 1988. The majority of those who came back had married Newfoundland girls and their wives and children accompanied them. They visited many of the former US military installation sites. On 18 September a US military reunion dinner, sponsored by the American Legion, was held at Fort Pepperrell Post #9 Headquarters. Over eighty former American servicemen and their wives attended, including eleven who first arrived in Newfoundland aboard the *Edmund B. Alexander*. The American Legion Fort Pepperrell Post #9 held an open house each day during the "Great '88 Soiree" so the public could look over the largest and most historic US military in Newfoundland display of photographs, documents, newspaper clippings and other items for the period 1940 to 1988. At the end of the event, all the material was donated to the Centre for Newfoundland Studies Archives, Memorial University of Newfoundland.

The American Legion, Department of Canada, was established in Newfoundland in 1941. When first organized in January 1941, the Post received its first charter (temporary) on 15 August of the same year and was named Fort Pepperrell #9. By the late fall of 1941, Post #9 had over 500 members. Eligibility requirements were that a member had to serve honourably in the US Forces in World War I from 6 April 1917 to 11 November 1918, or in World War II from 7 December 1941 onward. The first Post consisted of members from Fort Pepperrell and Argentia Naval Station. The permanent charter of the Post was awarded in the summer of 1941 by the National Executive Committee of the American Legion in the United States.

Top: Part of the 1,000 historic items on display at the American Legion Post #9 depicting the US military in Newfoundland and Labrador from 1940 to 1988.

Centre: Part of the scene of the American Legion Fort Pepperrell Post #9 reunion dinner, 18 September 1988.

Bottom left: Headquarters of the American Legion Fort Pepperrell Post #9, Building 1133, taken in the spring of 1987. The building is located in the White Hills which once formed part of the Fort Pepperrell property.

Bottom right:Advertisement in the American Legion Magazine, May 1988. The magazine has a circulation of 6.5 million.

YANK! COME BACK TO NEWFOUNDLAND
"THE GREAT '88 SOIREE"

The Biggest and Longest Party Ever Held— from May to October 1988

Every City and Town in Newfoundland and Labrador will be celebrating 100 years of Municipal Government with special activities. Visit the places where you served in the U.S. Armed Forces from 1941-1987, and reunite with family and old friends. See the many new changes in Newfoundland's way of life and fun, and compare it to the 'good old days'. Plan to return to Newfoundland this summer and participate in the 22 week long party of fun. Enjoy the special events at the American Legion Post 9 Headquarters, St. John's, during Regatta Week. Let us tell you all about the "Great '88 Soiree". For information, write or call **Great '88 Soiree, c/o American Legion, Fort Pepperrell Post 9, P.O. Box 9476, St. John's, Newfoundland, Canada, A1A 2Y4. Telephone Toll Free 1-800-563-NFLD.**

The Post was a most active one during the years of its existence in Newfoundland and still is. In 1953 the members of the Post applied to the Department of Canada Executive to have the name changed to Chilton-Tower Post #9.

The youngest son of S/Sgt. Earl Chilton died at Fort Pepperrell in 1951. M/Sgt. Porter B. Tower, who was the Base Sgt./Major at Fort Pepperrell from 1948 to 1951, also died in the same year. All members voted for a name change in memory of these two people. The National Executive approved the request and on 26 May 1953 the Post became known as Chilton-Tower Post #9. In 1954, the Post was awarded the National Commanders Cup for the highest paid-up membership of any American Legion Post in the United States and overseas. There are thirteen American Legion Posts throughout Canada. When it was known in late 1960 that Pepperrell Air Force Base was to close down and Argentia Naval Station was to be reduced in forces, it was decided on 8 March 1960 to petition the Department of Canada to have the Post's name changed back to Fort Pepperrell Post #9. The amendment to the Post's charter was approved on 23 December 1960.

Throughout the years of its existence, the American Legion Post has catered to the needs of its members and their families. Through close coordination with the American Consul in St. John's, and now Halifax, the Post has dealt with thousands of requests for information and action involving US veterans in benefits, burials, dependent inquiries and benefits and other personal matters. When Pepperrell Air Force Base closed in August 1961, and the American Consul's office was relocated to Halifax, Nova Scotia, the business of Fort Pepperrell Post #9 carried on as usual. Since 1941, the Post has been most active in community affairs within St. John's. Several donations were given to the orphanages and hospitals, as well as to such organizations as the Canadian National Institute for the Blind and the Janeway Children's Hospital. The American Legion youth programs have always been foremost. They sponsored minor baseball and softball teams, engaged in youth promotional activities through the Club, and started the children's identification program in Newfoundland.

All the Post #9 commanders served well during their terms of office. During the period 1952 to 1990 the following commanders served:

1952-53	Earl Chilton Jr.	1959-60	Richard C. Archer
1953-54	Robert W. Piatt	1960-61	John B. Elliott
1954-55	John Jacobs	1961-62	John F. Cornell
1955-56	Mac Brown	1962-63	Earl Richardson
1956-57	Louis F. McGilvary	1963-64	John B. Elliott
1957-58	Hampton Cayson	1964-68	(No Records)
1958-59	John N. Cardoulis	1969-70	John B. Elliott

1970-71	Billy Mercer	1980-81	Frank Jalosky
1971-72	John N. Cardoulis	1981-82	Dave Cook
1972-73	Fred Feuer	1982-83	John Mashburn
1973-74	John B. Elliott	1983-84	Bob Jergensen
1974-75	John N. Cardoulis	1984-85	Walter Gregoire
1975-76	John N. Cardoulis	1985-86	Jim West
1976-77	Russell Englund	1986-87	Hank Dopler
1977-78	Charlie Carr	1987-88	Robert Saul
1978-79	John Mashburn	1988-89	Robert Saul
1979-80	Ed Lupien	1989-90	Hank Dopler

The Ladies Auxiliary of Post #9, first established in 1958, was a strong organization long before they applied for and received their charter on 14 January 1972. The charter was officially acknowledged by the National Auxiliary Executive Committee, and signed by the National President, Bertha Parker. Charter members were:

Barbara Miller	Florence Hutchens
Mary Elliott	Elizabeth Jones
Joan Davis	Blanch Tower
Marjorie Kolbman	Phyllis Mashburn
Lora Johnson	Hazel Gross
Sylvia Adrens	Judy Mease
Doreen Cardoulis	Noreen Richardson
Betty Lupien	Jean Riddle
Olive Englund	Millicent Way
Gwen Harper	Ellen Gallager

Charter Member, Mrs. Blanch Tower, who now resides in Ontario, was the wife of the former M/Sgt. Porter B. Tower of Chilton-Tower Post. She was honoured in 1977 with a life membership in the Ladies Auxiliary.

A special five-day reunion was held at St. Lawrence from 29 June to 3 August 1988 for the remaining survivors of the ill-fated US *Pollux* and *Truxton* disaster of 19 February 1942. One hundred and forty US sailors died in the overall disaster. The reunion, sponsored by the Town of St. Lawrence, and spearheaded by Ena Farrell Edwards, was a great success. Over 2,000 townspeople and invited dignitaries attended, of which thirteen were some of the initial survivors. A fifty-car motorcade at St. Lawrence, with US and Newfoundland flags fluttering from every fender, welcomed the thirteen survivors and their wives.

As part of the American Legion "Yank! Come Back to Newfoundland" campaign, the US Ambassador to Canada and the Commanding General of the United States Marine Corps in Washington, approved the visit of the 2D Marine Aircraft Wing Band to St. John's on 5 and 6 July 1988. The fifty member band, commanded by Chief Warrant Officer 2, Roxanne M. Haskil, performed at Memorial Stadium to the delight of over 4,000

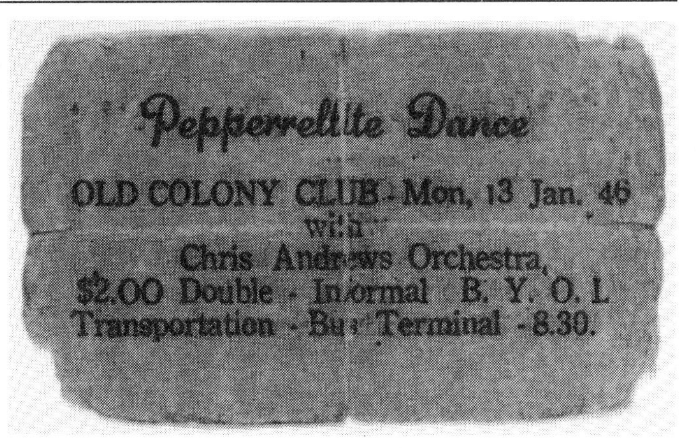

Pepperrellite Dance

OLD COLONY CLUB - Mon, 13 Jan. 46
with
Chris Andrews Orchestra,
$2.00 Double - Informal B. Y. O. L
Transportation - Bus Terminal - 8.30.

Ticket to Pepperrellite Dance at the Old Colony Club, 13 January 1946.

The 2D Marine Aircraft Band doing a precision march in the St. John's Memorial Stadium, 5 July 1988.

The 2D Marine Aircraft fifty-piece band out of Cherry Point, North Carolina performed on the aircraft ramp at Cherry Point, NC.

The first annual dinner of Fort Pepperrell Post #9 in the old Newfoundland Hotel, St. John's, 19 November 1941.

Official opening of Armed Forces Day, Pepperrell, 1952. Scene is to the right rear of the Base chapel. Military and civilian dignitaries are on the reviewing stand. The colours are being presented prior to the parade.

Newfoundland Lt. Governor, Sir Leonard Outerbridge, together with Col. Roy D. Butler, Pepperrell Base Commander, inspecting a Guard of Honour from the Air Police Squadron, prior to the start of Armed Forces Day celebrations , 1952.

spectators. Included in the program was a performance by the Signal Hill Tattoo and a fifty-man Honour Guard march by a contingent of RCMP, Royal Newfoundland Constabulary, St. John's Fire Department and Her Majesty's Penitentiary Guards. The US Marine Band stationed at Cherry Point, North Carolina arrived in their own DC-10 marine aircraft.

There are over 600 former US military personnel who served in Newfoundland and Labrador since 1940 now residing in Newfoundland. There are over 200 in the St. John's area alone, of which seventy-five are American Legionnaires. It is closely estimated that over 100 are in the Stephenville-Corner Brook area alone, 100 in the Argentia-Placentia Bay area, and 100 in Labrador and many towns on the Island, such as Grand Falls and Gander. Over the years, the American Legion records reveal over 100 have died and are buried here. Two of the first contingent of US soldiers who arrived on the *Edmund B. Alexander* in 1941 reside in St. John's with their families. Each year, many more return in retirement and reside here with their Newfoundland wife and their family. Without fail, each will say, "It's nice to be back. We sure missed Newfoundland these past years."

Many hunting and fishing camps in Newfoundland and Labrador were opened and operated by the United States service personnel. Jubilee Lake was one operated by Pepperrell and was opened July 1960. Several were located throughout Newfoundland from 1941 to 1945: Deadman's Pond, near Gander; Bottom Brook Camp, The Pond and Camp 33 near Stephenville. Special Service Organizations (SSO) furnished all supplies, including hunting and fishing gear, as well as boats, motors, vehicles, snowshoes and skis. Special aircraft would take crews out to these locations, land on the nearest water source, and call back for them in a few days.

Camp 3-Rivers at Eagle River, Labrador, 1955 was a recreational area for Goose Bay officers and enlisted men of the USAF. Fishing and hunting were some of the major sports.

This American dollar, belonging to Munden Bray of St. John's, was signed in 1941 by over forty US soldiers in the Beer Garden aboard the Edmund B. Alexander.

This is the site of the St. John's STADAN observatory, operated by NASA and NRCC from 1962 to 1971. The large building on the right centre and the arrangements of steel antenna plates at the rear were used in the beginning by the STADAN Observatory. The buildings to the left are the NASA Tracking Station, opened in 1971.

At one time the American dollar was worth $1.10 in Newfoundland currency. All the American personnel were paid in American cash. The stores, clubs, dance halls, theatres, and other places of assembly in the St. John's, Stephenville and Placentia areas gladly paid the extra ten cents on the American dollar. The American serviceman, with his high pay, even in 1941-48, was able to benefit greatly from the exchange rate. It was no trouble in those days to buy American dollars at bargain prices. As each base closed, over the years all of the furniture, supplies, vehicles and other miscellaneous equipment was turned over to Canadian Crown Assets for disposal. One could purchase a jeep for only $250 or a Banana Wagon for $500. Some of the merchandise purchased through Crown Assets was worth thousands of dollars and sold by bid at very low prices.

Many Newfoundland civilians had special invitations in 1947 to a dance held at Torbay Airport in Hangar 4. The music was supplied by the forty-six piece band of "Les Brown and his Band of Renown," from New York. Many big time bands in the United States performed here at the USO building and at Argentia, Gander, Ernest Harmon and Goose Air Base. Charlie Barnette and his orchestra, sponsored by the USAF at Pepperrell in 1953, played at the Drill Hall in Buckmaster's Field to raise funds for the St. John's Memorial Stadium.

During the initial launching of the first fourteen Apollo series of spaceships from Cape Kennedy, Florida, the National Aeronautical Space Administration (NASA) mission control at Houston, Texas, determined that even with twelve Satellite Tracking Stations in the US and around the world, there was one blind spot not covered, which resulted in a loss of the satellite for several hours tracking time while in orbit. The blind spot was over the Island of Newfoundland and immediate surrounding area. In 1970 NASA was given permission to build a satellite tracking station on a three-acre site at Shoe Cove, Newfoundland. A portable unit in the Bahamas was disassembled and reassembled at Shoe Cove, The elaborate facility opened in the fall of 1971. There were sixty American technicians on the tracking station and seventeen Newfoundland civilians. Mr. Gary Knight, who worked at the tracking station as a logistics employee and the longest employed by NASA, was one of the seventeen Newfoundlanders directly involved and assigned to the NASA Space Program. Mr. Knight was awarded a medallion by NASA for his service. The Shoe Cove facility played an important part in the tracking of the remaining Apollo space missions, including the first and second Skylab spaceships. On 15 July 1975 both the US Apollo and Russian Soyux spaceships blasted off into space to link up with one another on 17 July 1975 in the unprecedented two-day interstellar mission. The Shoe Cove tracking station was continuously used until 1976, by which time more elaborate and up-dated equipment had been put into operation by NASA and had eliminated the blind spot over Newfoundland. Except for electronic equipment, all the facilities at Shoe Cove were turned over to the Government of Newfoundland in 1977. Today it is used as a Lions Club Recreational facility. Very few people ever knew of Newfoundand's role in the Space Program for over fifteen years. In 1962, both NASA in the United States and the National Research Council of Canada established an observatory at Shoe Cove, known as the St. John's STADAN Station. The purpose of this installation was to study the stars and planets and the orbit of the earth, and to gather information for the proposed Apollo mission, which was to follow years later. When NASA assembled the satellite tracking station at Shoe Cove in 1971, both the observatory work and spaceship tracking was a combined operation.

The US Federal Bureau of Investigation (FBI) was very active in Newfoundland as far back as 19 December 1941. The first FBI men from Washington, D.C. arrived at St. John's aboard the *North Gaspé*, a private yacht out of Brooklyn, New York. There was at least one agent assigned to all US Army bases in Newfoundland during the war years, and several at Headquarters, Fort Pepperrell. After the war they were replaced by agents of the Office of Special Investigation (OSI). The FBI and OSI agents worked closely with the police authorities in St. John's and elsewhere.

US military personnel participated in many of the various parades held in St. John's and other locations in Newfoundland and Labrador, where they were stationed. After the war, their

When the USO opened in Corner Brook in 1944, the USAAF Band from Stephenville played for the gala dance event.

Above: Fort Pepperrell started its own dairy farm in 1946. Here, M/Sgt. Ed Rekkemper, NCOIC, proudly views the prize bull. Over thirty-two milk cows provided fresh milk daily to the mess halls and officers' and enlisted men's quarters. The dairy farm was operated until 1953, when a mechanical cow (pasteurized powdered milk) was put into operation. The cost of the mechanical cow was over $25,000.

The dining room of Ye Olde Brigus Tea Room, Brigus, Newfoundland, 1944. This was a favourite restaurant for visiting American servicemen.

participation was especially noted in Argentia, Stephenville and St. John's. The Army, Navy and Air Force Bands were in great demand at such functions as Armistice Day, Memorial Day, Battle of Britain Day, Santa Claus parades and many other special events. The US Army Air Corps Band and later the USAF 596th Band were star attractions at local dances and other functions.

From 1940 to 1945 the Newfoundland Railway carried thousands of tons of freight and thousands of passengers from St. John's to Argentia for the contractors and the US Navy. To facilitate the heavy freight load on the twenty-one miles of track between Placentia Junction and Argentia, the rail was changed by the Americans in 1942 from fifty to seventy-five pound rail. The complete railway was owned and operated by the Newfoundland Railway. Due to the shortage of railway stock to take care of the US military demand, the United States contributed over 100 flatcars, boxcars, tankers and passenger cars, so the railway could cope with the rapid increase of passengers and cargo movement across the Island.

In 1947, military personnel were allowed to bring their private vehicles on the base and they could purchase military gasoline for 10¢ a gallon, and a quart of oil for 5¢. On weekends, those without cars could check out military vehicles such as jeeps and carry-alls and go where they wanted on three-day passes. Over fifty or more vehicles would be checked out on weekends alone. The operation was called "R & R" (Recreation and Reconnaissance). Military personnel could invite any civilians to ride in the vehicle, providing he listed their names in advance on the application for an R & R vehicle.

21,940 Canadian Forces personnel participated in the Korean War from 25 June 1950 to 31 January 1955. Over 500 Newfoundlanders were a part of this force, and fought side by side with the American Forces. Several Newfoundlanders were killed in action. Newfoundlanders who enlisted in the United States also served in the Vietnam War from 22 December 1961 to 7 May 1975, and information from a local Vietnam veteran indicates there were Newfoundland casualties while they fought along with the American forces.

At Pleasantville today, only five of the original American contractor's temporary buildings remain and are still in use. They are T-851 on east White Hills Road, the original "Morris Hall," now used by the Department of Lands and Forestry; T-951 on Churchill Avenue, the original Base Chapel, now used by the Department of Education Film Library; T-930 on The Boulevard, the original Special Services Building, now used as the Royal Canadian Legion Headquarters Branch 56; and T-723, east of the Janeway Hospital, the original home of M/Sgt. Rekkemper, the dairy barn supervisor.

The home is now occupied as a residence on Newfoundland Drive; building T-943, behind building 910, is the United Sail Works.

When Fort Pepperrell officially closed on 11 August 1961, a caretaker force of about forty civilians under the command of Captain Billy Shannon of the USAF looked after the USAF interests until a Federal-Provincial Board decided what to do in reference to disposal. On 14 December 1961, the authority of the caretaker force was transferred to HMCS *Avalon*, Buckmaster's Field, St. John's. On 4 January 1962, orders were issued by the Canadian Naval Commander, in St. John's, of HMCS *Avalon*, appointing Mr. Harold Ellis as the Officer-in-Charge of Fort Pepperrell. All civilians were transferred to the Canadian Department of National Defence.

Plans for a monument to the tragedy of the Arrow Airlines DC-8 crash at Gander on 12 December 1985 are well under way. "Samaritan Place," as it will be called, will be a living monument to the 248 American soldiers and eight civilian crew members who died in the fiery crash on that fateful morning. "Samaritan Place," estimated to cost $30 million, is a project of the Gander Masonic Memorial Complex committee. The facility will be designed to accommodate 200 apartment units for those who were denied the benefit of retirement. Six of these apartments will be specially designed for the family and friends of the deceased US soldiers, whenever they visit the complex. In addition, there is proposed fifty beds for seniors who are in need of chronic care. In total, there will be 256 units, the same as the number of US soldiers who lost their lives in the crash of the Arrow Air jet. The main feature of the complex will be the central core and rotunda which has been designed specifically in memory of the soldiers. The rotunda will bear 36 wall panels, each carrying a flag of a state representing the men and women who died on the flight. The rotunda will also have a large non-denominational chapel, various recreational facilities, and a community centre. The setting of "Samaritan Place" overlooking Gander Lake will be exquisite.

Each character alive today, Newfoundland or American, who is part of this history could tell his or her own story of events and association with one another. The wars of the past five decades may be over, but there is still an uneasy feeling among many. The world has never reached the ultimate in peace. If a Third World War were to happen, everything would be very different from what we experienced in Newfoundland during World War Two. As one Air Force pilot recently said to the author, "Our generation has passed through the propeller age, the jet age and we are now in the push button atomic age. What we, in peaceful countries, had in machinery, aircraft, weapons, electric equipment, ships and manpower, over the last fifty years, would today be mostly redundant."

Father McNamara celebrating mass on Easter Sunday 1942 in an improved chapel on Signal Hill for members of the 24th Coast Artillery.

Easter dinner in a blackout at the mess hall of "B" Battery, Signal Hill, 1942.

Soldiers enjoying a coffee break in the PX coffee shop, 1942. Note the large poster portraying the "Good Soldier."

Sgt. Leo Marquis loading cows aboard a C-54 transport plane at Prince Edward Island, 1948. The cows were carefully selected from PEI stock and flown to Pepperrell's dairy farm.

Epilogue

It is most interesting to note that on 4 May 1965 all the area known as Fort Pepperrell was renamed "Pleasantville." This was the original name of the north shore area of Quidi Vidi Lake back in 1914, where the Royal Newfoundland Regiment trained prior to deployment to England for service in the First World War. The original military history of Pleasantville, later Fort Pepperrell, and again Pleasantville in 1965, is one of the great legends in the history of Newfoundland. The last battle of the "Seven Years War" was fought in 1762 on the Pepperrell and Quidi Vidi site between the French and the British forces. The British defeated the French in that year when they overtook the French garrison on Signal Hill. Thus, the military skirmishes in Newfoundland ceased in the eighteenth century.

The twentieth century, however, brought a renewal of military rule to the shores of Quidi Vidi. The Americans in 1940 began construction on a huge US Army base. Although the Anglo-American Lease was for ninety-nine years, the United States relinquished all claim to the land, and the buildings and facilities, built upon it, as of 14 December 1961, after twenty-one years of occupancy.

The appointed Canadian military authority over all the property at Fort Pepperrell as of 4 January 1962, was the Royal Canadian Naval Command in St. John's, HMCS *Avalon*. In 1940, in an agreement between Canada and England, the Canadians agreed to accept the responsibility of defending the harbour at St. John's. HMCS *Avalon* was established on 13 June 1941, and was commissioned in September of that year. Their headquarters was first in the Newfoundland Hotel. In July 1941 construction began on a Royal Canadian Naval Site in St. John's, however, it was never fully occupied. Later the HMCS *Avalon* moved into Buckmaster's Field in St. John's and remained there until 10 December 1961, when they moved into Building 313 at Fort Pepperrell. HMCS *Avalon* maintained the responsibility of all of the Fort Pepperrell facilities until 1964, when the buildings and facilities of Pepperrell were turned over to both the federal and provincial governments.

In 1949, when the colony of Newfoundland became the tenth Province of Canada, the Canadian Tri-Service Headquarters, which consisted of personnel from the permanent force in the Navy, Army and Air Force, was also located at Buckmaster's Field. The Royal Canadian Naval Reserve was first introduced in 1949, and on 20 September 1949 was officially commissioned as a Royal Canadian Naval Reserve Division in Newfoundland and was known as HMCS *Cabot*. They moved into Building 314 at Pepperrell in June 1963. HMCS *Cabot* maintains the responsibility of the Canadian Naval Reserve Division in St. John's. In 1963, the Canadian Minister of National Defence announced that, with the exception of the HMCS *Cabot* and some reserved Canadian Army units, defence activities and establishments at Pepperrell would close down. On 9 April 1964 HMCS *Avalon* was paid off, and as previously noted, all buildings that were not to be used by the federal government were to be turned over to the Province. One year later the name "Fort Pepperrell" was replaced by "Pleasantville" and so ended the American era.

HMCS *Avalon* was designated the responsibility of organizing and summer training of all Canadian Army, Air and Naval Cadet Corps and Leagues in the immediate St. John's area. The organizations under HMCS *Avalon* for training are:

514 Kinsmen Royal Canadian Air Cadet Squadron
Fort Townsend 166 Royal Air Cadet Squadron
Terra Nova #3 Royal Canadian Sea Cadet Corps
2515 St. John's Royal Canadian Army Cadet Corps
508 Caribou Royal Canadian Army Cadet Squadron
515 North Atlantic Royal Canadian Air Cadet Squadron
510 Lions Royal Canadian Air Cadet Squadron
93 Navy League, Fort Pepperrell, Cadet Squadron

The Tri-Service Headquarters at Buckmaster's Field in St. John's officially closed in 1962, and all Canadian military activities moved to Pepperrell. The Royal Newfoundland Regiment, also formed in 1949, moved to Pepperrell in the same year. Their headquarters is located in Building 311. The Royal Newfoundland Reserve Regiment consists of:

1st Battalion - The Royal Newfoundland Regiment
36th Service Battalion
56th Field Engineering Squadron
2nd Battalion Army Reserve District Headquarters

In 1966 a large drill hall for use of all the Canadian Forces was built at Pleasantville by the Department of National Defence, in an area behind the former USAF Officers' Club.

NLCC #93 Fort Pepperrell Ships Company, 1989. Several members are absent from this photo.

US Army troops parade on the drill field at Fort Pepperrell, 4 July 1942. Note the 1776 dress of two soldiers carrying the flags.

The so-called Tri-Service Headquarters, formerly at Buckmaster's Field, became the Canadian Forces Station (CFS) in 1962, when the Canadian Department of National Defence set up its operation at Pleasantville. The CFS has a contingency of Army, Air Force and Navy. The station has a force, including reserves, of approximately 800 personnel. Organizations that make up the Canadian Forces Station are:

727 Communications Squadron (Regular Force)
728 Communications Squadron (St. John's)
Royal Newfoundland Reserve Regiment

Other organizations with military backgrounds, located at Pleasantville today are:

American Legion, Fort Pepperrell Post #9
Provincial Command Royal Canadian Legion
Royal Canadian Legion, Branch #56
RCAF 150 North Atlantic Wing Association

Pleasantville is also the headquarters for "B" Division of the Royal Canadian Mounted Police in Newfoundland and Labrador.

The only building on the main base that has been torn down is the old base theatre. All other buildings are occupied by either the federal or provincial government, or have been sold to private enterprise, such as the base exchange building. The West Motor Court is occupied by the Canadian Coast Guard. All the barracks in the 800 block have been converted to sixteen-apartment units. Base headquarters and the Air War Room are used by the Canadian Forces Station as its headquarters. Most of the East Motor Court is occupied by the Cabot Institute, a provincial government Technical College for various trades. The temporary barracks and mess hall on the White Hills are gone, and the rifle range area has been developed into a huge housing project. All the buildings have been painted pastel colours of yellow, blue and green. The three Pepperrell gates (entrances) are still used, however, no guards and no guard gates are present. The area is completely public. The parade grounds have been turned into a number of sports facilities. The base hospital is now the renowned Dr. Charles Janeway Children's Hospital. Several additions have been added to the building. Overall, the whole former base area has been well preserved and maintained. Most of the 400 and 500 block apartments are occupied by Canadian military personnel and RCMP officers. Uniformed personnel are still a common sight on the old US military reservation.

In Argentia, the US Navy vacated the north side of the base in 1972, which was the original US Navy installation and consolidated all activity into the former McAndrew Air Force Base area on the south side. The Naval base is still most active. Part of the marginal wharf is maintained by the Navy, and the remainder is utilized by

—(USAF-NEAC Photo)

OVER 100 YEARS OF PEPPERRELL EMPLOYMENT are represented by these ten civilians—each of whose 10-year-and-over career here is a success story. Seated (l. to r.) are Pepperrell's Deputy Commander, Col. Ward W. Harker; James T. Vinicombe—the "oldest" (Oct., 1940) base employee who has risen from assnt. messenger to fiscal accountant; and Lt. Col. James C. Petersen, staff personnel officer. Standing (l. to r.) are: Fred G. Clark, NEAC Civilian Personnel Officer from Junior Clerk; Matthew P. Downey, Mobile Equip Maintenance Foreman from Ignitionman; Susan Downton, Laundry Supervisor from Forelady; George LeShanз, Commissary Administrative Assistant from Sr. Clerk; Carmel Kemp, Pepperrell Commander's Secretary from Jr. Typist; Phyllis M. Dunn, Telephone Supervisor from Jr. Tele. Oper.; George Rabbitts, Property & Supply Officer (M.C.G.) from Clerk; Harold F. Ellis, Construction & Maintenance Superintendent from Checker; Melvin C. Hollett Civilian Administrative Officer from Jr. Clerk

The lowering of the US flag at Pepperrell Air Force Base, 11 August 1961. In twenty years of operations the base provided a lifetime of memories.

PLUGS FIRST, LAST LINES AT PEPPERRELL

Miss Phyllis Dunn of St. John's has a unique distinction. She was the telephone operator who took the first call which came into the switchboard at Pepperrell Airforce Base when that base was established 18 years ago. And Miss Dunn Thursday at midnight took the last call to come into the base. Pepperrell's switchboard closed down. The last call to come into the base was made by base public relations officer Gordon Fillier who telephoned a few minutes before midnight to ask for the main gate. Miss Dunn, switchboard supervisor, recalled that the first call came into the base just a few seconds after the switchboard was set into operation. She said: "The call was for the officers' club." How many calls has Miss Dunn taken during her career as operator at the base? Said she: "Countless thousands." At peak periods about 25,000 calls a day came into the base switchboard. The team of operators were understandably "extremely busy," said Miss Dunn. She has often answered 5,000 calls in a single day. She has spoken with persons in every state of the United States and in most countries of the world since she became an operator at the base which will be phased out completely before the end of the year. Miss Dunn said: "It was a sad thing to receive the very last call . . I loved working at the base . . everyone has been so kind to me." Miss Dunn said she will now take a holiday and will then seek another job . . . "as a telephone operator."

the federal and provincial governments. A new daily ferry service has been established between Sydney, Nova Scotia and Argentia. Very few of the old buildings on the north side are occupied today. All the hangars and other large support structures have deteriorated badly over the years. The runways were closed in 1973 and the area no longer exists as an airport. Many new structures have been constructed on the present Navy side, including a ten-storey, 500-man modern dormitory and office complex with a luxurious setting, built in 1958. Top security is the order of the day throughout the US Navy installation. Argentia is the only US military base left in Newfoundland in conjunction with the original Anglo-American Lend Lease agreement.

Ernest Harmon Air Force Base officially closed in 1976. It is now part of Stephenville and Stephenville Airport. All the buildings and facilities were turned over to a Federal-Provincial Board in 1977. The federal government, under its Department of Transport, took control of all the airport facilities, with the exception of some of the flight line hangars. The provincial government took control of all other property and established a governing board called the Harmon Corporation in 1966. The duties of the Harmon Corporation were to maintain the facilities and endeavour to have new industry established, with the intent of utilizing the old military structures. At the same time, some of the buildings and facilities were turned over to the Town of Stephenville. In 1970 a new, modern linerboard mill was constructed by Port Harmon. The mill is still operational today, but under control of a different company. Port Harmon was upgraded as a marine terminal in 1970 and is a most modern facility today.

The two large dormitories known as the "Harmon Hilton" are used today by the Community College of St. George's as dormitories and classrooms for this provincial educational facility. The Harmon Corporation released several buildings for provincial use. The base hospital was renovated and today serves the community as the Sir Thomas Roddickton Hospital. All the former officers' and NCO quarters have been turned into public housing. Many of the barracks buildings have been converted into apartment units. The base chapel is operated by the Pentecostal faith and the base theatre by Famous Players Inc. Hangar 1 has been completely renovated and is now the town's arena. The modern school building is used as a large commercial store. Several small and semi-permanent buildings have been demolished to make way for new construction. Many of the hangars and flight line buildings, including the Black Hangar, are used today to house many new industries and other private enterprises. Gradually, the Town of Stephenville has taken over the control of many of the existing structures on the former base.

Texas Towers, like the one in this photo, were used by the Air Defence Command anchored several miles off the US shoreline to maintain a twenty-four-hour surveillance of the seaward approach to Newfoundland, Canada and the United States. The tower was virtually a floating AC&W Station. One of these were positioned off the west coast of Newfoundland on training manoeuvres for a short time in 1962.

There were several ammunition storage igloos constructed on Signal Hill from 1941 to 1945. This photo shows one of the original ones in the area of the Queen's Battery. One was constructed in 1943 of solid concrete and was removed from the site in the early '70s. A large ammunition dump was located at Portugal Cove. Twelve underground ammunition storage igloos still remain on the White Hills. Some are in private use today as hydroponic gardens and general storage areas.

Memorial Day Parade, 1950. Photo shows the American contingent marching east on Duckworth Street, St. John's.

In 1987 the Harmon Corporation was dissolved and the control of all facilities under its jurisdiction came under the Newfoundland and Labrador Housing Corporation. Like Pepperrell, many of the white asbestos-shingled buildings have been painted pastel shades of yellow, blue and green. The area, like Pepperrell, has been well maintained except for some hangars and flight line support buildings, for which no commercial use has been found. The Town of Stephenville has extended its boundary to include a lot of Harmon area. The military rail line to Stephenville Crossing is still in existence, but seldom used, especially since the complete Newfoundland Railway line across the province ceased to exist as of 1989. The former US Air Force Base looks as beautiful as ever today, no matter how much it has been divided by various provincial, municipal and private concerns.

Gander Airfield, the busiest of all US Army Air Corps installations during the war years, is now part of the large Town of Gander, and the airport is known as Gander International Airport. Most all of the small, temporary wartime buildings of both the RCAF and the USAAF are gone. While some hangars have been demolished, others have been upgraded over the years and are in full use today. Some of the old USAAF buildings that were maintained since 1942-43 are in use today. The old base hospital is now the RCAF station administration offices. The old heating plant has had its third storey removed and renovated, and is in use by the Department of Transport. A local and international airport terminal building was constructed in 1959, and is located on the former American side of the base.

The Canadian Forces moved into Gander early after 1949 and in 1951 took over control of the new American-built AC&W station at Gander. The thirty-acre site is known today as Canadian Forces Base, being elevated to that status from Canadian Forces Station in 1964. The base has a total military and civilian complement of 450 personnel.

Goose Air Base closed on 1 October 1976, after being under the operational control of the USAF Strategic Air Command for the last fifteen years. All USAF structures and facilities remained intact when they were turned over to the federal Government of Canada and provincial government. Over the years, some of the buildings have become a part of the township of Happy Valley. Like Harmon, many of the permanent facilities were purchased by private enterprise, however, the main part of the airport was maintained, and is still operated by, the Canadian Department of Transport. The base hospital was taken over by the Department of Health and the facility is now the Melville Hospital. The Officers' Club is now the American Forces Military Air Command (MAC) Club, and the NCO Club is not occupied at all.

Although Goose Bay remained dormant for a few years after closing, except for local and trans-atlantic commercial flights, plans were being made by the Canadian Government to use the existing facilities. In 1984 several NATO countries became interested in utilizing the former base as a training centre for low flying jet aircraft manoeuvres. Today the British Royal Air Force, Royal Netherlands Air Force, West German Air Force and the United States Air Force are all actively engaged in these exercises. The decision by the North Atlantic Treaty Organization to officially name Goose Bay as the training centre is still under negotiations. Already, the construction of additional hangars and other support facilities has been completed by some NATO country groups. There is no doubt Goose Bay will be as important a base in the future as it was in the past.

The Long Line Repeater Stations across the Island of Newfoundland have long disappeared, with the exception of the one at Whitbourne. This is being used by the township to house the local fire department. The transmission lines laid by the Signal Corps forty-eight years ago have been replaced with more modern equipment. There is little or no evidence of the five USAAF radar sites at Sandy Cove, Torbay, Cape Bonavista, St. Bride's and Allan's Island. The US military camp at Colinet is now a large housing project, and all that remains of Camp 4 on Salmonier Line is the old staff house. The Direction Finding station on the "Ridge" at Wesleyville in 1942 is not even known today to most of the youth of the town. All traces of the military weather stations and transmitter sites on the Island and in Labrador are gone. Only shells of buildings remain at Red Cliff, but one can still go down in one of the US Army Artillery bunkers.

Many of the costly Aircraft Control and Warning Sites and Gap Filler Stations in Newfoundland and Labrador are no longer there, except for the one built at Gander in 1951 and now operated by the RCAF, and some of the old USAF AC&W sites in Labrador and Baffin Island are in use today by the Canadian Government. The new underwater transmission cables from Thule Air Base in Greenland to Cape Dyer on the east coast of Baffin Island, and then overland to Hampden, White Bay and Wild Cove, Deer Lake in Newfoundland, and undersea again to Cap de Roches on the Quebec shore provide a direct link of communication from Thule Air Base to NORAD headquarters in Colorado Springs, Colorado. This, along with the installation of more updated and sophisticated aircraft and missile detection equipment, caused a number of AC&W stations and sites to become redundant as of 1961.

The two former USO buildings, in St. John's and Corner Brook, were sold in 1947 and are being used by private enterprise. The US Army Dock in St. John's is jointly used by the St. John's Port Authority and Ultramar Oil of Canada. The one at Port Harmon is now a marine terminal. The US Memorial Hospital at St. Lawrence has been reduced in activity to clinic status. Just about all the temporary buildings at Torbay Airport are gone and replaced by new commercial airline structures. The gym is now the Provincial Recreation Centre and all hangars are used by commercial airline companies.

Except for a record of past historic events, nothing remains at Signal Hill to recall the activity of the US Coast Artillery and Infantry. The total area today is a most beautiful tourist attraction, especially around Cabot Tower. Where the Coast Artillery barracks and other buildings were located, a large National Historic Park Museum has been constructed. It contains a narrative history and some photographs of the American occupation of the site during the Second World War years.

The changes that have taken place during the last five decades have almost eliminated the past US military presence here. From the day when it all began, up to 1976 when the most active American military days ceased, the course of Newfoundland's history changed. One cannot say "it's like a dream," because the reality of their presence throughout our province remains. Each year it is so pleasant to welcome back more former US servicemen and women, who will inevitably say, "The land has sure changed over the years, but the hospitality of the Newfoundland people has never changed. It's so wonderful to come back again."

The last all-military wedding at Argentia Naval Station was celebrated 29 June 1973, when Chief Petty Officer Robert (Bob) Joergensen married Hazel Snow of Harbour Grace at the Base Chapel. The gala event, held in the Chief's Club, featured a flowing fountain of champagne. Bob retired from the service in 1972 and lives in Mount Pearl with his family.

The first contingent of US Army troops loading aboard the US Transport for departure from Newfoundland, August 1945.

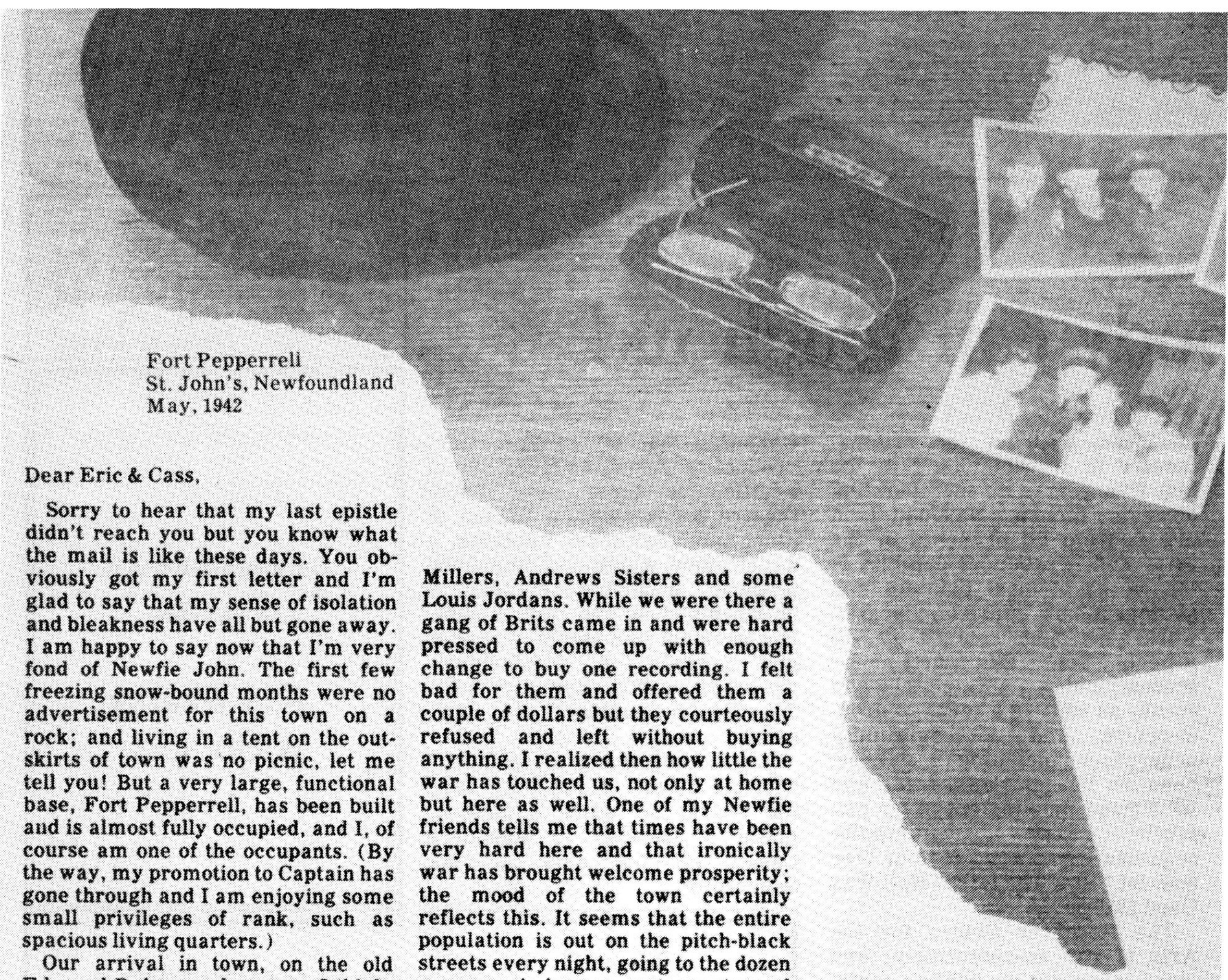

Fort Pepperrell
St. John's, Newfoundland
May, 1942

Dear Eric & Cass,

Sorry to hear that my last epistle didn't reach you but you know what the mail is like these days. You obviously got my first letter and I'm glad to say that my sense of isolation and bleakness have all but gone away. I am happy to say now that I'm very fond of Newfie John. The first few freezing snow-bound months were no advertisement for this town on a rock; and living in a tent on the outskirts of town was no picnic, let me tell you! But a very large, functional base, Fort Pepperrell, has been built and is almost fully occupied, and I, of course am one of the occupants. (By the way, my promotion to Captain has gone through and I am enjoying some small privileges of rank, such as spacious living quarters.)

Our arrival in town, on the old Edmund B. has made waves, I think. Although there are a lot of Cannucks and Brits off the convoys here, our numbers have caused the military population (the civilian as well) to double in size. The town is jumping! Several of the natives (no, not Eskimos) have remarked that the place is overrun with servicemen, and aren't things lively!

I was surprised and delighted to find all the latest run movies are shown here. Last Saturday, I saw Joan Crawford's new movie. I went with a swell girl called Eileen who I met at the Caribou Hut. It's an entertainment centre, and sort of a home away from home for all us servicemen (from the States and other places too). The other day Jeff and I (you remember Jeff, I brought him home on a furlough last year) went to the local music store to stock up on the latest recordings. We came away with 12 or 15 scorchers, Glen Millers, Andrews Sisters and some Louis Jordans. While we were there a gang of Brits came in and were hard pressed to come up with enough change to buy one recording. I felt bad for them and offered them a couple of dollars but they courteously refused and left without buying anything. I realized then how little the war has touched us, not only at home but here as well. One of my Newfie friends tells me that times have been very hard here and that ironically war has brought welcome prosperity; the mood of the town certainly reflects this. It seems that the entire population is out on the pitch-black streets every night, going to the dozen or so movie houses, restaurants, and dance halls, where 10-piece bands (local and imported) play well into the night as the energetic dancers beg for more.

The locals are some of the most open and friendly people I have ever encountered. Emotions are close to the surface and the sense of comaraderie overwhelming. I have made friends here that I know will be friends for life. Nothing is too much trouble and hospitality is a way of life and taken very seriously. The only gaff I've made so far is being late for a dinner engagement. My friend invited me to tea on Sunday about a month ago, but didn't specify what hour to come, so when I got off duty at six I waited until after dinner to sip a little Newfie tea. To my consternation I discovered that they call dinner 'tea'. Dinner (or tea) was ruined but everyone was so gracious, it became no more than an amusing incident.

Cass, you would love it here, I wish you could visit. Maybe after this whole thing is over, you two could come up for a holiday. When we shipped out from New York we all thought we were bound for the Arctic. Instead we find ourselves in a lively, progressive town that I have a feeling I'll be coming back to again and again.

Anyway, I really have to go now. Frank Sinatra is heading up the bill at a big USO show down on the base. Eileen and I are looking forward to a great evening. (It's funny to think that I would never have gotten to see him back home.)

So long, see you soon. And Cass, you look out for that old sodbuster Eric, don't let him spoil that great pitching arm of his.

All my love, as always.
Chuck

Wartime letter of an American serviceman in St. John's compiled by Janis Spence from material researched by the cast for the stage production of Making Time with the Yanks.

The photographs reproduced here and on the following two pages show civilian personnel working at Pepperrell Air Force Base in 1953. Do you recognize anyone?

Appendices

US Military Locations ~ Newfoundland

Legend

(1) Air Force Bases
(2) U.S. Naval Station
(3) U.S.A.F.A.C. & W. Site
(4) U.S. Army Repeater Station
(5) U.S. Infantry Site
(6) U.S. Coastal Artillery Site
(7) Meeting location Roosevelt and Churchill 10 Aug. 1941
(8) Transmitter Sites
(9) Radar Site
(10) Direction Finding Station
(11) U.S. Army Camp Site
(12) C.O.T.C. and U.S.A.F. Cable Buildings

St. Anthony (3)
Lascie (3)
Hampden (12)
Sandy Cove (9)
FOGO ISLAND
Deer Lake (12)
Corner Brook (4)
Howley (4)
Port au Port (5)
Jerry's Nose (3)
Harmon AFB (1) (4)
Gander (1) (4) (3)
Stephenville (5) (6)
Stephenville Xing (4)
Millertown Junction (4)
Grand Falls (4)
Wesleyville (10) (8)
Cape Bonavista (9)
Elliston (3)
Shoal Hr. (4)
Arnold's Cove (5)
St. Andrews (11)
Table Mountain (8)
Cape St. Francis (5)
Pouch Cove (5)
Flat Rock (5) (6)
Middle Cove (5)
Outer Cove (5)
Logy Bay (5)
Torbay (9)
Robin Hood Bay (5) (6)
Red Cliff (3) (5) (6)
Signal Hill (5) (6)
Cape Spear (5) (6)
White Hills (6)
Pepperrell AFB (1) (3)
Torbay Airport (1)
Snelgrove (8)
ST. JOHN'S
Argentia Navy (2)
McAndrew AFB (1)
Whitbourne (4)
Isaac's Head (7)
Placentia (5)
Colinet (11)
Fox Island - Dunville (5) (6)
Allan's Island (9)
St. Bride's (9)

APPENDIX A

US Military Locations in Newfoundland and Labrador 1940-1990

US Air Force and Army Bases
- Pepperrell AFB, St. John's
- McAndrew AFB, Placentia
- Gander Air Base, Gander
- Torbay Airport Air Base, St. John's
- Ernest Harmon AFB, Stephenville
- Goose Bay Air Base, Labrador

US Naval Station
- Argentia Naval Air Station

US Army Coast Artillery Sites
- White Hills, Pepperrell, St. John's
- George's Pond, Signal Hill, St. John's
- Red Cliff, St. John's
- Placentia
- St. George's
- Stephenville
- Port-au-Port
- Flatrock
- Cape Spear
- Robin Hood Bay
- Fox Island, Dunville

US Army Infantry Sites (Look-out Towers)
- Logy Bay
- Middle Cove
- Torbay
- Outer Cove
- Flatrock
- Robin Hood Bay
- Pouch Cove
- Placentia
- Stephenville
- Arnold's Cove
- Port-au-Port
- Cape St. Francis
- Fox Island, Dunville

US Army Camp Sites
- Colinet
- Camp 4, Salmonier Line
- Camp Alexander, St. John's
- Camp Morris, Stephenville
- Camp 33, St. George's

US Transmitter Sites
- Snelgrove
- St. John's
- Table Mountain
- St. Andrew's
- Wesleyville

US Direction Finding Stations
- Wesleyville
- Gander
- Harmon
- Goose Bay
- Argentia

NASA and NRCC Observatory Station and NASA Satellite Tracking Station
- Shoe Cove
- Pouch Cove

US Army Signal Corps Repeater Stations
- Fort Pepperrell
- Whitbourne
- Shoal Harbour
- Gander
- Grand Falls
- Millertown Junction
- Howley
- Corner Brook
- Stephenville Crossing
- Stephenville

US Military Locations
~ Labrador

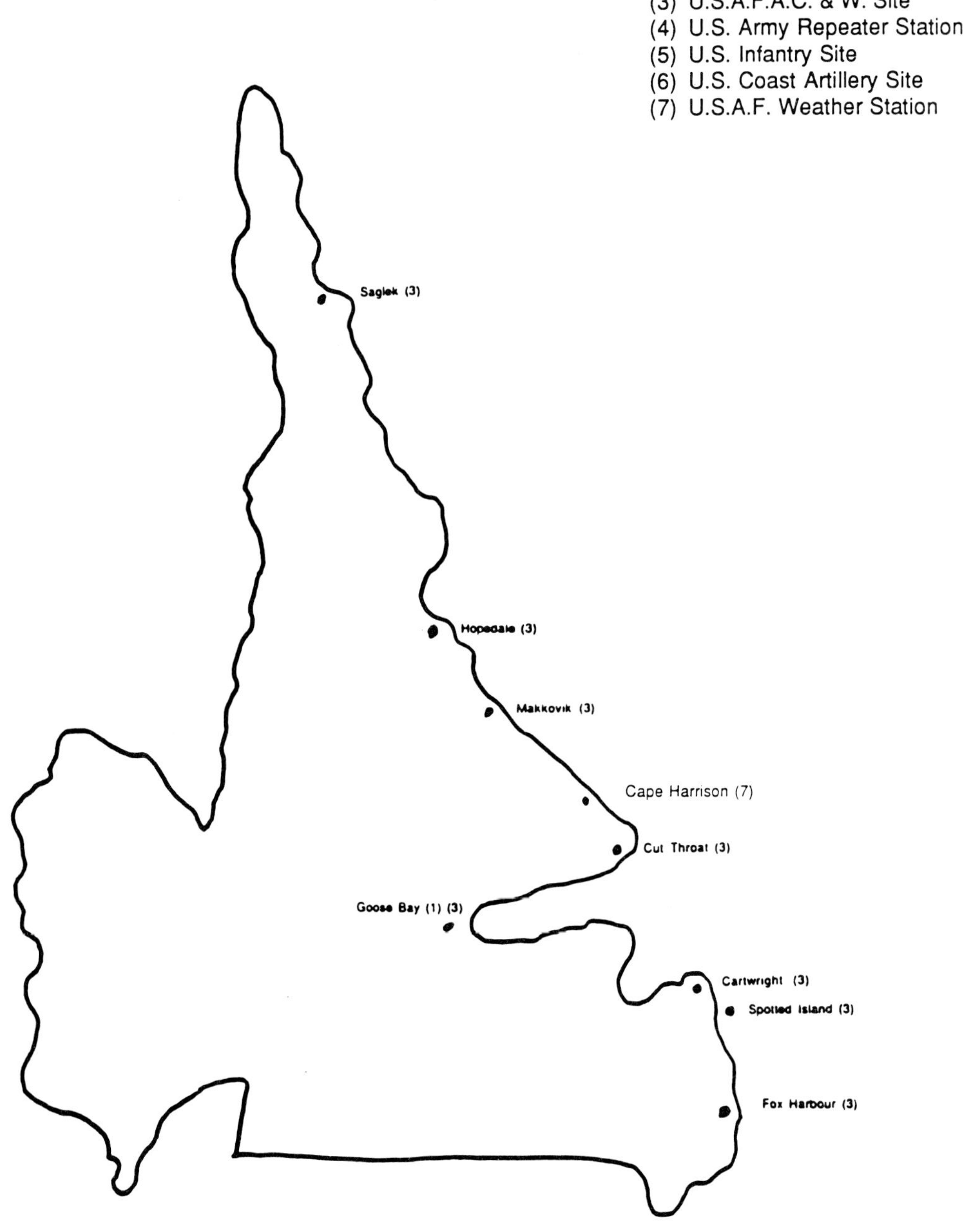

US Army Radar Sites
- Sandy Cove, Fogo Island
- Cape Bonavista
- Torbay
- St. Bride's
- Allan's Island

USAF and COTC Cable Buildings
- Hampden
- Deer Lake

Current (1990) Locations of US Military in Newfoundland
- Argentia Naval Station
- Gander
- Goose Bay, Labrador

USAF Weather Station
- Cape Harrison, Labrador

Master Control AC&W Stations, USAF and Navy
- Argentia Naval Station
- Ernest Harmon AFB, Stephenville
- Goose Bay, Labrador
- Thule Air Force Base, Greenland

US Air Force AC&W Stations and Radar Gap Filler Stations
- Red Cliff, St. John's
- McAndrew AFB, Placentia
- Elliston Ridge, Bonavista Bay
- LaScie, White Bay
- Gander
- Stephenville
- St. Anthony
- Jerry's Nose, Port-au-Port
- Goose Bay, Labrador
- Cartwright, Labrador
- Hopedale, Labrador
- Sagalek, Labrador
- Cape Makkovik, Labrador
- Cut Throat Island, Labrador
- Spotted Island, Labrador
- Fox Harbour, Labrador

Other US Military Locations with Headquarters at Pepperrell Air Force Base 1941 to 1961
- Indian House Lake, Quebec (weather station)
- Fort Chimo, Quebec (weather station)
- Frobisher Bay, Baffin Island (AC&W station)
- Ungava, Ungava Bay (weather station)
- Mingan, Quebec (weather station)
- Resolution Island, Resolute Bay (AC&W station)
- Thule Air Force Base, Greenland
- Sonderstrom Air Base, Greenland
- Narsarssuak Air Base, Greenland
- Padloping Island, NWT (weather station)
- T-3 Floating Ice Station, Thule, Greenland (experimental Arctic laboratory)

APPENDIX B

Chronological History of Argentia Naval Air Station

1940 — Site selected in Placentia Bay. Originally called Little Placentia, it was renamed Argentia.

October 13. USS *Bowditch* arrived Argentia with US Corps of Engineers and civilian hydrographic and surveyor personnel.

1941 — January 19. USS *Richard Peck* arrived Argentia with 1,500 US construction workers and engineers.

January 25. First all-military troops arrived at Argentia when 120 US Marines came ashore.

January 28. First construction work began at Argentia.

February 13. US Marines officially raised first US Flag.

May 15. First Navy sea plane patrol squadron arrived.

July 15. Naval Operating Base Argentia officially commissioned. First Commanding Officer was Captain J.A. Morgan.

July 30. Vice Admiral A.L. Bristol, Commander Task Force 24, arrived on USS *Prairie* to take command.

August 10. Site of rendezvous between US President Roosevelt and British Prime Minister Churchill.

August 28. Naval Air Station Argentia officially commissioned.

September 17. Five US destroyers from Argentia met a fifty ship merchant marine convoy and escorted it across the Atlantic.

1942 — September. US Navy Wing 7 and US Army Air Corps unit were stationed at Argentia.

October. The 17th Naval Construction Battalion (Cee-Bees) arrived at Argentia. The Battalion was part of the 10th Construction Regiment, consisting of 1,049 personnel. They assumed construction responsibilities and base maintenance from the US civilian construction companies.

November. British Navy established a shop maintenance base at Argentia for their escort destroyers.

1943/45 — Over 10,000 US Naval, Marine, Seabees and Army troops in Argentia. Thousands of ships and war planes passed through Argentia en route to European Theatre of War.

1945	May. European war over. Japanese war continued until August 1945.
	November. Most of US Army troops and artillery moved out of Argentia.
1947	Argentia Naval activity increased; personnel and equipment increased.
1948	Approximately 7,000 Naval and Marine troops at Argentia.
1950	Korean War alert. Military personnel increased.
1953	Argentia assigned support to Northeast Air Command.
	Approximately 8,500 Naval, Marine and Air Force troops in Argentia.
1955	Adjacent McAndrew Air Force Base closed. Navy took over some facilities.
1956	Use of airstrip at Argentia curtailed.
	Korean War over; Naval and Marine personnel reduced.
1958	Argentia assigned support to 64th Air Division and NORAD.
1969	US Naval and Marine troops reduced to 3,000.
1971	US Naval and Marine troops reduced to 1,000.
1972	Deactivated buildings and facilities on north side turned over to Government of Canada.
1973	US Air Force at Argentia relinquished all command to US Navy. Argentia Airfield closed.
	US Marine troops vacated Argentia. Only Navy personnel remained.
1975	On north side, federal government occupied 20% of all buildings and facilities, provincial government 40% and private enterprise 40%.
1989	US Navy still occupies south side of Base.
1990	Argentia today one of the most modern US facilities in Navy.
1990	Many former US servicemen live in surrounding area of Argentia at Placentia, Fox Harbour, Freshwater, Jerseyside and Dunville.

APPENDIX C

Chronological History of Fort Pepperrell, St. John's

1940	Site location for Pepperrell chosen. Field work began by US Corps of Engineers.
1941	Construction of US military site began on the north side of Quidi Vidi, Signal Hill and US Infantry and Coast Artillery sites.
	UST *Edmund B. Alexander* arrived in St. John's (29 January). 977 US Army troops aboard.
	Col. Maurice D. Welty, US Army, First Commanding Officer Newfoundland Base Command.
	Camp Alexander opened on Carpasian Road.
1942	Joint use of Torbay Airport; RCAF and US Army Air Corps (USAAC).
	Construction of US Army Dock began. Fully operational in 1943.
1943/44	Over 5,000 US Army troops at Pepperrell and immediate area.
1945	European war over; Japanese war continued until August 1945.
	Most of US Army troops moved out of Pepperrell and other areas.
1946	Fort Pepperrell designated as US Army Air Corps Base.
1947	Major construction began at Pepperrell for conversion to USAF facility.
	Most of remaining US Army troops transferred to US Air Force.
	Fort Pepperrell transferred to US Air Force and renamed Pepperrell Air Force Base.
	RCAF deactivated military side of Torbay Airport.
	US Air Force took over operation of Torbay Airport (military side).
1948	Approximately 2,000 USAF personnel at Pepperrell and Torbay Airport.
1950	Korean War Alert increased number of personnel at Pepperrell.
	Pepperrell expanded to construct White Hills area.

1950	Pepperrell redesignated as Headquarters Northeast Air Command (NEAC).
1953	Over 4,000 US Air Force personnel at Pepperrell and Torbay.
	Major construction program on White Hills.
	Korean War over; personnel contingent reduced to 3,000.
1957	Northeast Air Command began to decrease its activity.
1958	NEAC activities transferred to 64th Air Division Command.
1960	Phase-down of Pepperrell Air Force Base began.
1961	August 11. Pepperrell Air Force Base closed, caretaker force only.
1962	Pepperrell and all facilities turned over to federal and provincial governments.
1963	Major construction program to change occupancy of most buildings.
1963	75% of all former military barracks buildings converted to16-unit apartments.
1965	May 4. Pepperrell Air Force Base area changed back to "Pleasantville."
1975	Canadian government occupied 30% of all buildings and facilities, provincial government 50% and private enterprise 20%.
1988	Over 500 US ex-servicemen returned to Pepperrell for reunion.
1990	Many former US serviceman still live in St. John's and near vicinity.
	Only reference to former Pepperrell Air Force Base is Fort Pepperrell Post #9 of the American Legion, and the Canadian Fort Pepperrell Naval Cadet Corps #93.

APPENDIX D

Chronological History of US Army Infantry Locations

Logy Bay - Outer Cove - Middle Cove - Torbay - Flatrock - Pouch Cove - Robin Hood Bay - Fort McAndrew - Arnold's Cove - Placentia - Port au Port - Stephenville - Cape St. Francis.

1941	Designated locations for US military look-out towers.
	Construction began on towers and security facilities.
	Facilities occupied by the 3rd US Army Infantry personnel.
	Over 200 US Army personnel assigned on rotation basis.
1942	Some troops assigned to man look-out towers at Fort McAndrew and Stephenville.
1945	European war over; Japanese war continued until August 1945.
	Facilities deactivated; US Army troops reassigned to USA and overseas.
1947	Buildings and other facilities removed from site.
1990	Some evidence of infantry locations still in existence.

APPENDIX E

Chronological History of US Army Coast Artillery Locations

George's Pond, Signal Hill - Red Cliff - Fort McAndrew - Stephenville - Flatrock - Cape Spear - Robin Hood Bay - White Hills - St. George's - Port au Port - Stephenville

1941	Designated locations for US Military Coast Artillery.
	Construction began for stationed US Army Artillery Troops at Signal Hill.
	Facilities occupied by US Army 24th Coast Artillery and 62nd Anti Aircraft Artillery.
1942	Some troops assigned to man Artillery at Fort McAndrew and Stephenville area.
1943/44	Over 4,500 US Army personnel assigned to Coast Artillery units.
1945	European war over; Japanese war continued until August 1945.
	US Army troops returned to United States.
1946	Facilities deactivated.
1947	Buildings, artillery and most other facilities removed from site.
1990	Two of the original Coast Artillery buildings are still on site at Signal Hill, as are some of the ammunition storage bunkers remain at Signal Hill, at base of Southside Hills, the White Hills and Argentia.
	Some of the old gun emplacements are still to be found at most locations.

APPENDIX F

US Spending in Newfoundland, 1941-1961

From 1941 to August 1961 the United States government spent over $150 million on the construction of buildings and areas at Fort Pepperrell, Signal Hill, Torbay, Camp Alexander, Army Dock and others in the immediate St. John's area. The following is a published US Air Force paper on the financial contribution of Pepperrell to the local economy of St. John's and surrounding areas for the fiscal year of 1958 (31 March 1958 to 1 April 1959):

A. Financial contribution to the local economy (1958):

Appropriated Funds	
Civilian Payroll	$4,965,925
Supplies and Services	4,117,591
Non Appropriated Funds	
Officers' and NCO Club—Airman's Club—Post Exchange	
Civilian Payroll	451,647
Supplies and Services	559,310
Personal Expenditures of US military	
Rentals	1,347,900
Purchases (Local)	774,227
Total	$12,216,600
Total in 1957	13,663,204

B. Base Acreage and Value:
Under the 99-year lease renegotiated by the US with Canada, Pepperrell AFB encompassed a total area of 1,635 acres with an improvement value of $31.2 million. The original cost of Fort Pepperrell up to 1944 was $25 million. The current value in 1958 was over $100 million.

Roads Maintained

Roads	Miles Paved	Miles Unpaved	Total
On Base	5.3	6.6	11.9
Off Base*	5.2	5.0	10.2
	10.5	11.6	22.1

C. Utilities:
An average of $198,000 per month was paid the Newfoundland Light & Power Company and United Towns Electric Company for an average of 944,250 kilowatt hours of electricity. At the same time, a monthly average of 18,670 million gallons of water is purchased from

These are Torbay Road, Logy Bay Road and The Boulevard.

the Municipality of St. John's at a cost of $1,807 per month. Pepperrell also maintained 38,250 (linear feet) of sewer lines; 46,434 linear feet of water lines; 76,165 linear feet of power lines and 14,820 linear feet of fuel lines; four fuel storage tanks; and a fire station consisting of forty civilians and three major items of fire fighting equipment. This was in addition to modern airport crash trucks and fully trained personnel at Torbay Airport and a Command Firefighting School on the White Hills.

D. Personnel Strength:

Civilian Personnel	1,264	Number of Families	820
Military Personnel	1,511	Total Dependants	2,302
Total	2,774		

F. Aircraft, Telephone, and Hospital Services:
There are 12 Aircraft assigned to Pepperrell's flying facility at RCAF Station, Torbay. These include, L-20 Reconnaissance planes, C-47 and C-54 Cargo aircraft, T-33 Jet Trainers and SA-16 Seamaster aircraft.

Telephone service is extensive. There are 1,700 telephone instruments installed on the Base. From these, an overall average of 750,000 calls are placed monthly.

A 40-bed hospital is maintained here. On an average, 29 beds are occupied each month. An average of 18 deliveries and 1,653 out-patient treatments are made each month.

The following is from *The Evening Telegram* files as prepared in May 1966:

About 500 rental units as indicated above are used by Naval personnel at Argentia. $75,000 monthly into the economy of Placentia, Freshwater, Jerseyside and Dunville.

There are approximately 25,000 Americans stationed at US Bases in Newfoundland and Labrador. Say that each spends $5.00 per month in the Province, it would equal $1,500,000 per year.

The Bases buy over $1,000,000 a year of products such as eggs, bread, soft drinks, beer, etc. from Newfoundland.

In 1959 Americans living in St. John's spent $1,023,000 in rentals.

The Bases purchase approximately $6,000,000 worth of meat, vegetables, and other products each year in New Brunswick, Prince Edward Island and Nova Scotia.

All employees at Marconi, located at Pepperrell Air Force Base work on a $3,000,000 a year USAF contract repairing electrical equipment for US Bases in this area of Canada.

By the end of the fiscal year 1959-60 (31 March 1960) the civilian payroll for Pepperrell workers and those at satellite (Radar) sites was $5,236,429, the military payroll was $6,751,467, supplies purchased cost $1,932,459 and contract service $1,070,978. Total rentals paid by Americans living in rented housing in and around St. John's for 1959 was $1,203,210.

APPENDIX G

Base Commanders, 1941-1961

Down through the years, from the days of the Newfoundland Base Command in 1941 to the 64th Air Division up to 1961, the following commanders served:

Col. Maurice D. Welty	January 1941 to July 1941
Brig. General H.W. Harms	July 1941 to September 1941
Maj. General G.G. Brant	September 1941 to October 1941
Lieut. Col. J.J. Yeats	November 1941 to December 1941
Col. Paul W. Starling	December 1941 to January 1943
Brig. General John B. Brooke	January 1943 to October 1944
Brig. General Samual Connell	October 1944 to December 1945
Col. Albert E. Warren	December 1945 to January 1946
Brig. General Caleb V. Haynes	January 1946 to July 1948

Pepperrell Commanders since then:

Col. Ira D. Snyder	January 1948 to August 1949
Col. James G. Pratt	August 1949 to July 1950
Col. Roy D. Butler	* to July 1953
Col. Richard W. Fellows	July 1953 to June 1954
Brig. General William H. Wise	June 1954 to August 1955
Col. Graeme S. Bond	August 1955 to August 1957
Col. Richard E. Decker	August 1957 to July 15, 1960

Not Base Commanders, but in Charge of Designated Commands:

Maj. Gen. Lyman P. Whitten Newfoundland Base Command & NEAC	July 1949 to July 1953
Lt. Gen. C.E. Myers NEAC	July 1953 to July 1954
Lt. Gen. Gleen O. Barcus NEAC	July 1954 to April 1957
Col. C.W. McColphin 65 Air Div., ADC	April 1957 to July 1958
Brig. Gen. F.W. Terrell 64 Air Division, ADC	July 1958 to June 1960

**The date Col. Butler replaced Col. Pratt is unknown.*

APPENDIX H

US Military Personnel Strength

Army — Air Force — Navy — Marines — CeeBees
All Locations 1940 to 1961

★ The Tour of Duty in Newfoundland for all US Military from 1941 (after Pearl Harbour) to 1945 was for the duration of the war.

★ The Tour of Duty from 1945 onward was:

single or married without dependents—2 years
married with dependents—3 years

This meant a complete turnover of personnel every three years after 1945.

★ Some requested, and were granted, extended Tours of Duty.

	Prior to 1946	Assigned Personnel 1941-1961
Pepperrell	5,000	4,000
Harmon	4,000	3,000
Argentia	10,000	8,000
McAndrew	3,000	1,000
Goose Bay	3,000	3,500
Gander	4,000	1,000
AC&W Stations	-	3,500
Rep. Stations	400	200
Artillery	2,000	-
Infantry	4,000	-
	*34,900	**23,200

Personnel - 1946 to 1952 = 12,000 x 3-year rotation = 36,000
Personnel - 1952 to 1955 = 23,000 x 3-year rotation = 69,000
Personnel - 1955 to 1958 = 23,000 x 3-year rotation = 69,000
Personnel - 1958 to 1961 = 23,000 x 3-year rotation = 69,000

**Figures are based upon known information gathered over the years, and from ex-US servicemen who first came to Newfoundland on the UST* Alexander, *served extended Tours of Duty and have resided in Newfoundland ever since.*

*** The above figure of 23,200, multiplied by the Tour of Duty years, is well over 100,000 troops.*

APPENDIX I

Chronological History of Gander Air Base

1936	Construction began as a commercial airport.
1938	First aircraft landed, piloted by Capt. Doug Fraser.
1940	Canadian Government began construction of RCAF base.
1941	RCAF Station Gander in operation.
	US Signal Corps, AACS, 8th Weather Squadron arrived.
	US Army Air Corps began construction of American side.
1941/44	Over 4,000 US Army Air Corps troops assigned to Gander.
1942	First contingent of 150 US Army Air Corps arrived.
1943/44	Thousands of American and Canadian Bombers, Fighters, Freighters and Troop Carrying Aircraft stopped over at Gander en route to European Theatre of War.
1945	European war over in May; Japanese war continued to August.
	Thousands of American and Canadian military aircraft passed through en route to home from European Theatres Gander used primarily for commercial aircraft.
1947	Reduction of US troops as military mission diminished.
1948	American side of Gander deactivated. Buildings turned over to Newfoundland government.
1949	Gander named "Crossroad of the World."
1949/50	Many American military buildings demolished or renovated.
1950	Gander designated as International Airport.
1950/60	Contingent of US Air Force personnel assigned at Gander.
1954	Gander incorporated as a town.
1985	December 12. Crash of Arrow Airlines, killing 248 soldiers of the 3rd Battalion, 502nd Infantry, 101st Airborne Division, en route home from UN Contingency Force duty on the occupied West Bank of the Sinai. The soldiers were stationed at Fort Campbell, Kentucky.
1990	Gander today is one of the busiest international airports in North America. Small contingent of US Air Force personnel still at Gander. Several US ex-servicemen live in Gander and surrounding towns.

APPENDIX J

Chronological History of Ernest Harmon Air Force Base

1940	Military base location in St. George's Bay area, adjacent to Stephenville.
1941	Construction of US Military site began. Named Stephenville Air Base under Newfoundland Base Command.
	Harmon Field officially named.
	Over 500 US Army (Corps of Engineers) assigned on construction.
1942	23 March: contingent of US Infantry and Artillery transferred from Pepperrell to Harmon.
	Over 700 US Army Air Corps personnel arrived and set up Camp Morris Tent City. More troops quickly followed.
	Branch of Newfoundland Railway extended from Stephenville Crossing to Harmon. This was built by the US Army.
1942/44	Over 4,000 US Army and US Army Air Corps personnel stationed at Harmon and immediate areas. Thousands of US military aircraft passed through Harmon en route to European Theatre of War.
1943	Air strips ready for military aircraft.
1945	European war over-Japanese war continued until August 1945.
1945/46	Most of US Army troops transferred out of Harmon. U.S. Army Corps maintained airport and facilities.
1947	Major construction began on conversion to US Air Force base.
	Mission of Ernest Harmon changed to include additional defence responsibilities.
1948	Ernest Harmon transferred to United States Air Force and renamed Ernest Harmon Air Force Base.
	Approximately 2,000 USAF personnel stationed at Harmon.
1950	Korean War Alert increased number of personnel at Harmon.
	Construction of Port Harmon begins.
1953	Redesignated part of the Northeast Air Command.
	Facilities expanded at Harmon. Military personnel strengthened to 5,000. Port Harmon extended.
1956	Korean War over-personnel contingent reduced.

1957 New construction began at Harmon Air Force Base.

1958 NEAC phase out-Harmon under 64th Air Division and Strategic Air Command.

1965 Deactivation of Harmon Air Force Base began.

1966 Harmon closed; all US military facilities at Harmon turned over to Federal and Provincial governments.

Control of all former base facilities came under Harmon Corporation. Operation of airport under Canadian Department of Transport.

1967 Beginning of conversion of many buildings into private business occupancies.

1970 Reconstruction began on Port Harmon as a Marine facility.

1985 Harmon Corporation held control of over 50% of former military facilities, 40% was private enterprise and 10% the Town of Stephenville.

1987 Interest controlled by Harmon Corporation taken over by Newfoundland and Labrador Housing Corporation and Town of Stephenville.

1990 Many former US servicemen live in Stephenville and surrounding area.

Airfield used today for commercial aircraft and known as Stephenville Airport.

APPENDIX K

Chronological History of Fort McAndrew

1940	Site chosen adjacent to Argentia in village area originally called Marquise.
1941	Construction of base began by US Corps of Engineers and American contractors.
	Newfoundland Base Command established.
	US Infantry and Coast Artillery assigned.
1942	First contingent of US Army personnel assigned.
	Road from Argentia to Holyrood constructed by Army.
1942/43	Over 3,000 US Army, Infantry and Artillery troops on base.
1945	European war over-Japanese war continued until August 1945.
	Most US Army troops left Fort McAndrew.
1946	Fort McAndrew designated as US Army Air Base using Argentia runway.
1947	Fort McAndrew transferred to US Air Force and renamed McAndrew Air Force Base.
1948	Approximately 1,000 US troops at McAndrew Air Force Base.
1950	Korean War Alert-personnel increased.
1953	McAndrew Air Force Base incorporated into Northeast Air Command.
1954	Deactivation of McAndrew Air Force Base began.
1955	US Navy took over all facilities on McAndrew area site.
1965	Remaining buildings and facilities on north side turned over to Canadian government.
1990	Many former US servicemen live in surrounding town areas such as Placentia, Freshwater, Dunville, Jerseyside and Fox Harbour.

APPENDIX L

Chronological History of US Army Signal Corps Direction Finding Stations

Wesleyville — Gander — Harmon — Goose Bay — Argentia

and Repeater Stations

Whitbourne — Shoal Harbour — Gander — Windsor — Millertown Junction Corner Brook — Howley — Stephenville Crossing — St. Andrew's

1940 Sites chosen as Military Communications Relay Stations.

1941 Construction began on locations by Signal Corps Company and Corps of Engineers.

1942 Communications Signal Lines installed across the province, St. John's to St. Andrew's.

Direction Finding Station and Radio Beacon installed at Wesleyville.

Each Repeater Station occupied by ten to fifteen US Army personnel, depending on size and workload of locations.

1943 Transmitters installed on Table Mountain and Snelgrove.

Second transmission cable installed from St. John's to St. Andrew's.

1945 European war over; Japanese war continued until August 1945.

1946 Deactivation of some Repeater Stations began.

Military installation at Wesleyville officially closed.

1947 Repeater Stations closed. Maintenance of some communications equipment at some locations maintained.

1949 Buildings at many locations were passed over to federal, CNT Communications, or sold to private enterprise.

1950 Military Communications Systems changed to AC&W Stations (Aircraft Control and Warning) Network.

1990 Of all these buildings, only one is still in use today, by private enterprise, at Whitbourne.

APPENDIX M

US Army-Newfoundland Long Line Systems

The US Army-Newfoundland Long Line System was comprised of the following stations and equipment:

★ Three circuit control terminal stations located as follows:

- Station "A" terminal located at Fort Pepperrell. Station "A" of the Newfoundland Long Line System was designated as the facility control station for the entire system.
- Station "G" terminal located at Gander.
- Station "M" terminal located at Harmon Field.

★ There were seven Repeater Stations known as "attended" stations which were located as follows:

- Station "B" located at Whitbourne.
- Station "E" located at Shoal Harbour.
- Station "G" located at Gander (combined control terminal and repeater station).
- Station "J" located at Grand Falls.
- Station "K1" located at Millertown Junction.
- Station "K2" located at Howley.
- Station "L" located at Corner Brook.

APPENDIX N

Chronological History of Goose Bay Air Force Base

1940/41	Construction of giant airport by Canadian government.
1942	RCAF officially took over in March, as RCAF Station Goose Bay.
	First contingent of US Army Air Corps personnel arrived.
	United States Army Air Corps began construction on American side.
	American Command of Goose Bay under USAAC, Manchester, New Hampshire.
1943/44	Over 2,500 American, 2,000 RCAF, 500 RAF, and over 700 civilians assigned to Goose Bay.
	Thousands of military bombers, fighters and troop carriers passed through Goose en route to European Theatre of War.
1945	European war over in May; Japanese war continued until August.
	Over 2,500 USAAF personnel assigned.
	Goose Bay was stopover for thousands of military aircraft and troops returning from European Theatre of War.
1946/47	Goose Bay named as strategic stopover trans-Atlantic flights.
1947	Goose Bay designated as a United States Air Force Base.
	Most of the RCAF area deactivated. Airport still under joint use.
	US part of Goose Bay transferred to United States Air Force.
1948	Approximately 3,500 military personnel at Goose Bay.
1949	Over 3,000 USAF personnel assigned.
1950	Korean War alert. US military personnel increased.
1953	Goose Bay incorporated into Northeast Air Command.
	Over 4,000 US military personnel stationed at Goose Bay.
	Beginning of reconstruction and expansion program for US and Canadian facilities at Goose.
	Goose Bay became strategic area in US defence program.
1956	Korean War over. Some military personnel reductions.
1960	Deactivation of Northeast Air Command began.

1961	Many buildings and areas in Goose Bay closed.
1962	USAF personnel at Goose Bay less than 1,000.
1976	Goose Bay AFB officially closed 1 October.
1976/87	USAF facilities gradually turned over to federal government.
1986	Goose Bay airport used primarily as fighter trainer base by national and international organizations, and private enterprise.
1990	Small USAF contingent still there to cater to US interests.
	Several former US servicemen work and live in Goose Bay and town of Happy Valley.

APPENDIX O

Chronological History of Radar Aircraft Control and Warning Stations (AC&W)

Sandy Cove — Cape Bonavista — Torbay — St. Bride's — Allan's Island — Red Cliff— Argentia — Goose Bay — Harmon — Gander — Elliston — St. Anthony — Cartwright — Hopedale — Sagalek — Resolution Island — Frobisher — Cut Throat — Fox Harbour — Makkovik — Spotted Island — La Scie — Jerry's Nose — Port au Port

1942/44	North Atlantic's first ground radar site operated at Sandy Cove, Fogo Island, followed by Cape Bonavista, Torbay, St. Bride's and Allan's Island.
1950	Sites chosen for Aircraft Control and Warning Stations.
	Construction began on locations.
	Korean War alert-Cold War with Russia.
1951	Occupancy began as each station was ready.
1952	Between 20 and 200 USAF personnel assigned depending on size and workload of each facility.
1955	Facilities expanded at many locations to include "Pine Tree."
1956	Korean War over-Cold War with Russia eased.
1959	Deactivation of most AC&W stations began.
1960/61	CNT took over operations buildings at some AC&W stations.
1965	Most buildings and facilities on stations sold or demolished.
1975	US Navy maintained a satellite communications centre at Red Cliff. Centre closed in 1982.
1990	Some operations at former sites operated by CNT or Canadian Military.

Credits

The photographs and documents reproduced throughout this book were kindly provided by the following individuals and organizations. The author expresses his gratitude. While every effort has been made to properly credit the owners of photographs, the author welcomes any additions and/or corrections to this list and will credit in future editions.

American Legion Magazine: 172 (bottom right)

American Legion, Post #9: 30 (bottom), 33, 36, 40 (top), 47 (top), 54, 57 (top), 59, 62, 63 (top left), 64, 90, 93 (top left & centre), 95 (top right & bottom right), 96, 99 (left), 100 (top two & bottom right), 123 (bottom right), 136 (top), 147 (bottom), 172 (top, centre & bottom left), 175 (bottom), 176 (centre & bottom), 184 (bottom right)

Barron, David: 145, 147 (centre)

Bray, Munden: 178 (top)

Burke, Frank: 22, 75 (centre), 106 (top two), 107, 109, 110, 111, 112, 113, 133, 134, 149 (top)

Butler, Frank: 123 (top left & centre)

Canada. Department of National Defence: 69, 71, 72 (bottom right), 115, 121 (top), 126

Canadian Forces Station Gander: 142

Canadian Red Cross: 161, 163 (top)

Cardoulis Family Collection: 119, 125 (bottom), 130 (top), 149 (centre right), 175 (top & centre), 200, 202

Centre for Newfoundland Studies, Queen Elizabeth II Library, Memorial University of Newfoundland: 20

Clancy, Tom: 196, 197, 198

Coady, Andrew: 51

Davis, Walter H.: 157 (centre), 162 (centre)

Dodd, Norman: 121 (bottom two), 123 (bottom left, top right), 128, 129, 137, 153, 169

Doyle, Mary: 155 (bottom left)

Dunn, Phyllis: 63 (centre right)

Edwards, Ena Farrell: 147 (top)

Englund, Russell: 57 (bottom), 58 (top three), 63 (bottom right), 157 (bottom left)

Fagan, Kevin: 47 (centre, left & right), 55 (centre), 149 (bottom), 156, 159 (right centre & bottom), 170 (top), 181 (top), 194 (right)

Falk Foto, C.: 80

Furlong, Wallace: 48

Evening Telegram, The (St. John's): 158, 170 (bottom two), 189, 195 (Kent Barrett photo)

Gander Masonic Memorial Complex, Inc.: 153 (bottom)

Gander Public Library: 72 (top)

Gander, Town of: 72 (centre & bottom left), 75 (left), 76, 77, 78, 79, 81, 82 (top)

Gibbons, Betty: 149 (centre left)